Comprehensive Guide to Animal Husbandry: From Production to Nutrition

Animal husbandry, a cornerstone of agricultural practices, encompasses a broad spectrum of knowledge and skills aimed at the responsible care, breeding, and management of domesticated animals. From the expansive landscapes of livestock farms to the intricate operations of dairy production, poultry farming, and apiculture, the art and science of animal husbandry play a pivotal role in ensuring food security, economic prosperity, and sustainable resource utilization.

The "Comprehensive Guide to Animal Husbandry: From Production to Nutrition" is a definitive resource that delves into the multifaceted world of animal husbandry, providing a holistic understanding of the interconnected disciplines that comprise this dynamic field. From the earliest origins of animal domestication to the latest advancements in breeding techniques and nutrition management, this guide offers readers a comprehensive exploration of practices that shape animal production, welfare, and the food systems that sustain humanity.

Spanning various species, production methods, and management strategies, this guide unfolds the complexities of animal husbandry through a series of well-structured sections. It elucidates the significance of genetic selection, reproductive technologies, and innovative breeding methods that underpin

1

the improvement of animal traits, yielding animals better suited for specific purposes.

The guide navigates through the intricacies of livestock management, touching on housing design, health maintenance, and the implementation of sustainable practices that prioritize animal welfare and environmental stewardship. As animal husbandry intertwines with advancements in technology, sections on meat and dairy production provide insights into modern practices that ensure safe and high-quality products reach consumers.

Nutrition, a cornerstone of optimal animal development, forms a core component of this guide. Readers will embark on a journey into the nutritional requirements of various species, the art of formulating balanced diets, and the incorporation of innovative feed sources that align with both animal well-being and resource efficiency.

Moreover, the guide takes readers into the fascinating realms of apiculture and poultry farming, shedding light on the delicate interplay between bees and agriculture, as well as the intricacies of managing poultry for egg and meat production.

Ultimately, this comprehensive guide serves as a vital resource for students, educators, researchers, and practitioners in animal husbandry. It unveils the critical role of animal husbandry in ensuring a sustainable and resilient food supply chain, while addressing ethical considerations, environmental impact, and the necessity for evolving practices in an ever-changing world.

As we embark on this journey through the multifaceted world of animal husbandry, let us uncover the timeless traditions, scientific breakthroughs, and innovations that collectively shape the path towards a more nourished and sustainable global community.

The Importance of Animal Husbandry in Agriculture and Food Production

Animal husbandry holds profound importance in the realm of agriculture and food production, playing a vital role in ensuring the sustainability, efficiency, and quality of our global food supply. This practice encompasses the care, breeding, management, and utilization of domesticated animals for various purposes, ranging from meat and dairy production to pollination and agricultural labor. The significance of animal husbandry reverberates through numerous dimensions:

1. Food Security and Nutrition: Animal husbandry contributes substantially to food security by providing a diverse range of protein-rich foods, including meat, poultry, eggs, and dairy products. These animal-derived foods are integral to balanced diets, offering essential nutrients such as proteins, vitamins, and minerals that are vital for human health and well-being.

2. Livelihoods and Economic Prosperity: For countless communities around the world, animal husbandry serves as a source of livelihood and economic stability. Farmers and livestock producers rely on animal husbandry to generate income, sustain rural economies, and support their families. The industry also generates employment opportunities across various sectors, from animal care to processing and marketing.

3. Agricultural Sustainability: Animal husbandry plays a role in sustainable agriculture by promoting resource efficiency and recycling. Animals, such as cattle and poultry, convert agricultural byproducts and waste into valuable products like meat, milk, and eggs. Additionally, livestock can contribute to nutrient cycling in agricultural systems, improving soil fertility and reducing the need for synthetic fertilizers.

4. Biodiversity and Ecosystem Services: Certain animal husbandry practices, like apiculture (beekeeping), support pollination of crops, enhancing agricultural productivity and biodiversity. Bees, for example, contribute to the reproduction of numerous plant species critical for food production. Moreover, pastoral systems that involve livestock grazing can contribute to maintaining open landscapes and preventing habitat degradation.

5. Genetic Diversity and Conservation: Animal husbandry efforts often include the preservation and conservation of rare or endangered livestock breeds, safeguarding genetic diversity within livestock populations. These diverse genetic resources can serve as valuable reservoirs for traits that are adaptable to changing environmental conditions and disease resistance.

6. Rural Development and Infrastructure: Investments in animal husbandry lead to the development of rural infrastructure, including veterinary services, research facilities, and transportation networks. These investments contribute to the overall development of rural communities and improve access to essential services.

7. Environmental Sustainability: Efforts to enhance animal husbandry practices can mitigate the environmental impact of livestock production. Sustainable practices focus on efficient resource use, waste management, reduced greenhouse gas emissions, and responsible land use to minimize the ecological footprint of animal agriculture.

In essence, animal husbandry bridges the gap between agricultural productivity, economic prosperity, and the nutritional needs of populations worldwide. Through careful management, responsible breeding, and the application of scientific knowledge, animal husbandry supports the balance between meeting the growing demand for food and preserving the health of our planet. As society navigates the challenges of feeding a growing global population while safeguarding natural resources, animal husbandry continues to hold a crucial place in shaping a sustainable and resilient future for agriculture and food production.

Historical Evolution of Animal Husbandry Practices

The historical evolution of animal husbandry practices traces back to the earliest stages of human civilization and has undergone significant transformations over millennia. From nomadic hunting and gathering to sophisticated modern farming systems, the development of animal husbandry reflects the intricate relationship between humans and domesticated animals. Here is an overview of the key stages in the historical evolution of animal husbandry practices:

1. Domestication and Early Agriculture (10,000 - 8,000 BCE):

- The transition from a nomadic, hunter-gatherer lifestyle to settled agriculture marked the beginning of animal husbandry.
- Humans domesticated animals like dogs, sheep, goats, cattle, and pigs, initially for companionship, protection, and eventually for their utility in providing food, labor, and materials.

2. Emergence of Agricultural Societies (8,000 - 2,000 BCE):

- Agricultural communities began to emerge, leading to more organized animal husbandry practices.
- Animals were primarily kept for food production, draft power, and clothing materials.
- Simple shelters and enclosures were built to house animals, facilitating management and protection.

3. Development of Livestock Breeds (2,000 BCE - 500 CE):

- Selective breeding practices were adopted to enhance desirable traits in animals.
- Different breeds of livestock were developed to suit specific purposes, such as milk production, meat, and labor.
- Herding and grazing systems were established to manage livestock efficiently.

4. Medieval and Renaissance Periods (500 - 1500 CE):

- Animal husbandry techniques were refined, and knowledge of animal care and breeding expanded.
- The medieval period witnessed advancements in veterinary medicine, as care for sick animals gained importance.
- Draft animals played a pivotal role in agricultural labor and transportation.

5. Industrial Revolution and Modernization (18th - 20th Century):

- The industrial revolution brought about significant changes in animal husbandry practices.
- Advances in mechanization and technology impacted farming and animal management.
- Selective breeding, artificial insemination, and improved veterinary care led to enhanced animal productivity.

6. Contemporary Animal Agriculture (20th Century - Present):

- Modern animal husbandry is characterized by intensive livestock production systems designed to meet the demands of a growing global population.
- Intensive breeding, optimized nutrition, and advanced healthcare practices have resulted in higher yields of meat, milk, and eggs.
- Sustainable practices, animal welfare concerns, and

technological innovations continue to shape the future of animal husbandry.

7. Sustainable and Ethical Considerations (21st Century and Beyond):

- Current discussions on animal husbandry encompass sustainability, animal welfare, ethical treatment, and environmental impact.
- Alternatives such as organic farming, agroecology, and pasture-based systems are gaining attention as ways to balance productivity with environmental stewardship.

The historical evolution of animal husbandry practices reflects the intricate interplay between cultural, technological, economic, and ecological factors. From humble beginnings of domestication to modern agricultural systems, the journey has been marked by adaptation, innovation, and a growing awareness of the need to harmonize the well-being of animals, humans, and the environment. As societies continue to evolve, the story of animal husbandry serves as a testament to the dynamic relationship between humans and the animals that have shaped our history and sustenance.

Overview of Animal Production Systems

Animal production systems encompass a range of methods and practices used to raise domesticated animals for various purposes, including food, fiber, labor, and companionship. These systems vary based on factors such as species, geographical location, cultural preferences, and technological advancements. Here's an overview of the main types of animal production systems:

1. Extensive/Traditional Systems:

- Extensive systems involve minimal intervention and rely on natural resources and local ecosystems.
- Animals graze freely in open pastures or rangelands, obtaining much of their nutrition from forage.
- Common in regions with ample land availability and lower population density.
- Promotes animal welfare and allows animals to exhibit natural behaviors.

2. Intensive Systems:

- Intensive systems aim to maximize productivity through controlled environments and close management.
- Animals are kept in confined spaces, such as feedlots or poultry houses, and receive formulated diets.
- Common in areas with high population density and limited land resources.
- Requires advanced management practices, nutrition, and disease control.

3. Semi-Intensive Systems:

- Semi-intensive systems combine elements of both extensive and intensive methods.
- Animals have access to both natural forage and supplementary feeds.
- Provides a balance between productivity and animal welfare.

4. Pastoral Systems:

- Predominantly found in arid and semi-arid regions, pastoral systems involve nomadic or semi-nomadic herding.
- Animals are moved between grazing areas to prevent overgrazing and allow natural vegetation regeneration.
- Cattle, sheep, and goats are common livestock in pastoral systems.

5. Agro-Pastoral Systems:

- Combines animal husbandry with crop production.
- Animals contribute to nutrient cycling through manure, and crop residues may be used as feed.
- Promotes resource efficiency and sustainable land use.

6. Mixed Farming Systems:

- In mixed farming systems, both crops and livestock are integrated on the same farm.
- Livestock provide manure for fertilizing crops, and crop residues serve as animal feed.
- Enhances farm diversification and resilience.

7. Industrial/Commercial Systems:

- Characterized by large-scale production, intensive management, and high technology adoption.

- Animals are raised for efficient meat, milk, or egg production.
- Involves specialized facilities and production methods.

8. Organic Systems:

- Organic animal production adheres to specific standards that prioritize animal welfare, natural behaviors, and environmental sustainability.
- Animals are raised without synthetic inputs such as antibiotics and growth hormones.
- Emphasizes pasture-based systems and access to outdoor areas.

9. Free-Range Systems:

- Animals have access to outdoor areas and natural environments.
- Promotes animal welfare and allows animals to engage in natural behaviors.

10. Aquaculture Systems: - Involves raising aquatic organisms, such as fish, shrimp, and mollusks, in controlled environments. - Various methods include pond systems, recirculating systems, and marine cages.

Each animal production system has its advantages, challenges, and implications for animal welfare, environmental impact, and food security. The choice of system often depends on factors such as geographical location, available resources, cultural practices, and consumer preferences. As the world seeks to balance the demands of a growing population with sustainability and ethical considerations, the evolution of animal production systems continues to adapt to changing needs and circumstances.

Intensive vs. Extensive Systems

Intensive and extensive animal production systems represent two distinct approaches to raising domesticated animals for various purposes, such as food production, fiber, or labor. These systems differ in terms of their management practices, resource utilization, animal welfare considerations, and environmental impact. Here's a comparison of intensive and extensive systems:

Intensive Systems:

- **Management:** Intensive systems involve closely controlled environments where animals are kept in confined spaces, such as feedlots, poultry houses, or indoor facilities.
- **Space Utilization:** Animals in intensive systems have limited space and may be housed in high-density conditions to maximize production efficiency.
- **Feeding:** Animals in intensive systems are often fed formulated diets that are precisely balanced for nutritional needs.
- **Productivity:** Intensive systems aim to achieve high levels of productivity and maximize output, such as meat, milk, or eggs, per unit of input.
- **Animal Welfare:** Concerns arise about animal welfare due to potential confinement, limited space, and reduced ability to exhibit natural behaviors.
- **Health Management:** Close monitoring and management are necessary to prevent disease outbreaks due to close quarters and high animal density.
- **Environmental Impact:** Intensive systems may lead to

concentrated waste accumulation and environmental challenges if not properly managed.
- **Examples:** Poultry houses, feedlots for cattle, indoor aquaculture systems.

Extensive Systems:

- **Management:** Extensive systems involve minimal intervention and rely on natural resources and ecosystems to support animal growth and health.
- **Space Utilization:** Animals in extensive systems have access to larger areas, such as pastures or rangelands, for grazing and foraging.
- **Feeding:** Animals often rely on natural forage, grasses, or available vegetation for a significant portion of their diet.
- **Productivity:** Productivity may be lower compared to intensive systems due to the reliance on natural resources and lesser input.
- **Animal Welfare:** Generally, animal welfare is considered better in extensive systems as animals have more space and freedom to engage in natural behaviors.
- **Health Management:** Animals in extensive systems are exposed to more natural conditions, reducing the risk of disease transmission and outbreaks.
- **Environmental Impact:** Extensive systems can promote sustainable land use and contribute to local ecosystems' health if managed properly.
- **Examples:** Grazing livestock on open pastures, free-range poultry, extensive fisheries.

In essence, the choice between intensive and extensive systems depends on various factors, including available resources, economic considerations, cultural practices, consumer demands, and sustainability goals. Both systems have their merits and challenges, and modern agricultural practices often seek to balance productivity with animal welfare, environmental stewardship, and ethical considerations.

Sustainable Livestock Farming Practices

Sustainable livestock farming practices aim to balance the needs of animal production with environmental preservation, animal welfare, and economic viability. These practices seek to minimize negative impacts on ecosystems, promote animal well-being, and contribute to long-term food security. Here are some key sustainable livestock farming practices:

1. Pasture-Based Grazing:

- Allow animals to graze on natural pastures, which promotes nutrient cycling, reduces feed costs, and enhances animal health.
- Rotational grazing prevents overgrazing, supports soil health, and improves forage quality.

2. Agroforestry:

- Integrating trees and livestock on the same land provides shade, shelter, and supplemental forage.
- Trees contribute to carbon sequestration, biodiversity, and microclimate regulation.

3. Integrated Crop-Livestock Systems:

- Integrate livestock with crop production, using manure as organic fertilizer and crop residues as animal feed.
- Enhances nutrient cycling, reduces waste, and promotes resource efficiency.

4. Managed Grazing Systems:

- Utilize holistic or planned grazing systems to mimic natural herd movement patterns.
- Improves forage quality, soil health, and prevents soil erosion.

5. Low-Stress Handling:

- Minimize stress during handling, transportation, and processing to enhance animal welfare and meat quality.

6. Genetic Selection for Adaptability:

- Choose livestock breeds that are adapted to local conditions and require fewer external inputs.
- Enhances resilience and reduces the need for medications and supplements.

7. Efficient Feed Conversion:

- Focus on balanced diets that maximize feed efficiency, reducing the environmental impact of feed production.

8. Waste Management:

- Implement proper waste management strategies to minimize the impact of manure on water quality and ecosystems.
- Consider composting, anaerobic digestion, or utilization of manure as fertilizer.

9. Animal Health Management:

- Prioritize preventive measures through vaccination, biosecurity, and natural remedies to minimize the need for antibiotics.
- Focus on animal well-being to reduce stress and the risk of disease.

10. Alternative Protein Sources: - Explore alternative protein sources such as insects or algae as feed ingredients, reducing reliance on resource-intensive feed crops.

11. Climate-Resilient Practices: - Implement practices that reduce greenhouse gas emissions, such as improving feed efficiency and adopting renewable energy sources.

12. Precision Livestock Farming: - Utilize technology and data to monitor animal health, behavior, and performance, leading to targeted interventions and reduced waste.

13. Collaborative Approaches: - Collaborate with researchers, government agencies, and local communities to develop and implement sustainable practices.

Sustainable livestock farming practices recognize the interconnectedness of ecological, social, and economic systems. They seek to ensure that animal production meets current needs without compromising the well-being of future generations, while also contributing to the overall health of the environment and the communities they support.

Principles of Animal Breeding

Animal breeding is a science and art that involves selecting and mating animals to perpetuate desirable traits in the offspring. It aims to improve animal genetics, enhance productivity, and achieve specific breeding goals. The principles of animal breeding guide the systematic and ethical development of animal populations. Here are the fundamental principles:

1. Selective Breeding:

- Select animals with desirable traits, such as high growth rates, disease resistance, or milk production, to become parents of the next generation.
- Favorable traits are inherited by offspring, leading to genetic improvement over time.

2. Heritability:

- Some traits have a higher heritability, meaning they are more influenced by genetics than environmental factors.
- Traits with high heritability respond well to selective breeding.

3. Genetic Variation:

- Genetic variation provides the foundation for selective breeding.
- A diverse gene pool allows for the identification and selection of superior individuals.

4. Breeding Value:

- Breeding value is an estimate of an individual's genetic merit for a specific trait.
- It helps in comparing animals' genetic potential and informs mating decisions.

5. Phenotypic and Genotypic Selection:

- Phenotypic selection relies on observable traits, while genotypic selection uses genetic information.
- Genomic selection involves using DNA markers to predict breeding values more accurately.

6. Inbreeding and Outbreeding:

- Inbreeding involves mating closely related animals, which can lead to the expression of undesirable recessive traits.
- Outbreeding involves mating less closely related animals, maintaining genetic diversity and reducing the risk of inbreeding depression.

7. Genetic Diversity:

- Genetic diversity is critical to prevent genetic diseases and maintain adaptability to changing environments.
- Overemphasis on specific traits can lead to loss of genetic diversity if not managed carefully.

8. Selection Index:

- A selection index combines multiple traits into a single value to facilitate breeding decisions.
- It considers the economic value of each trait and aims to achieve balanced genetic improvement.

9. Crossbreeding:

- Crossbreeding involves mating animals of different breeds or lines to harness heterosis or hybrid vigor.
- Hybrid offspring exhibit improved performance due

to the complementary genetic contributions from the parent breeds.

10. Record Keeping and Data Analysis: - Accurate and comprehensive record keeping is essential for evaluating breeding outcomes and making informed decisions. - Statistical methods and data analysis help assess the effectiveness of breeding programs.

11. Genetic and Environmental Interaction: - Genetic expression is influenced by environmental factors such as nutrition, management, and housing conditions. - Breeding aims to find the optimal balance between genetic potential and environmental conditions.

12. Long-Term Perspective: - Animal breeding is a long-term endeavor that requires consistent commitment to continuous improvement. - Goals and strategies may evolve over time based on changing industry demands and scientific advancements.

By applying these principles, animal breeders can make informed decisions that lead to the development of animal populations with enhanced genetic traits, improved productivity, and a better fit for specific production systems and market demands.

Genetic Selection and Improvement

Genetic selection and improvement are integral components of animal breeding programs aimed at enhancing desirable traits and genetic potential within a population of domesticated animals. Through systematic selection and mating strategies, breeders can achieve genetic progress over successive generations. Here's an overview of genetic selection and improvement processes:

1. Identifying Desirable Traits:

- Breeders identify traits that contribute to the overall productivity, health, and adaptability of animals.
- Traits can be economically important (e.g., growth rate, milk yield) or related to disease resistance, reproduction, or carcass quality.

2. Collecting Data and Phenotypic Information:

- Accurate and detailed data on individual animals' traits are collected over time.
- Phenotypic information includes measurements, observations, and performance records related to the identified traits.

3. Calculating Estimated Breeding Values (EBVs):

- Breeding values estimate an animal's genetic merit for specific traits.
- EBVs are calculated using statistical models that consider an animal's own performance, pedigree, and performance of relatives.

4. Genetic Evaluations and Selection Index:

- Genetic evaluations rank animals based on their estimated breeding values.
- A selection index combines multiple EBVs into a single value, considering economic values assigned to each trait.

5. Genetic Testing and Genomic Selection:

- Genetic testing uses DNA markers to predict an animal's genetic potential for various traits.
- Genomic selection enhances accuracy in predicting breeding values, especially in traits with low heritability.

6. Mating Strategies:

- Mating strategies are designed to combine desirable traits from different animals while avoiding undesirable traits.
- Inbreeding and outbreeding are managed to balance genetic diversity and avoid inbreeding depression.

7. Genetic Diversity and Balancing Selection:

- Maintaining genetic diversity is crucial to prevent loss of adaptive potential and susceptibility to diseases.
- Balancing selection involves managing traits to ensure genetic stability over time.

8. Monitoring and Feedback:

- Breeding programs are continuously monitored to assess progress and adapt to changing goals or market demands.
- Feedback from industry stakeholders and end-users informs breeding strategies.

9. Genetic Gains and Economic Impact:

- Over successive generations, genetic selection and improvement result in increased performance and productivity.
- The economic impact of genetic gains contributes to improved profitability and sustainability.

10. Long-Term Perspective: - Genetic improvement is a long-term process that requires patience and consistent adherence to breeding goals. - Achieving substantial genetic progress may take several generations of animals.

Genetic selection and improvement involve the strategic combination of science, technology, and practical expertise to enhance animal genetics in alignment with industry needs and consumer preferences. By focusing on key traits, utilizing advanced techniques, and considering genetic diversity, breeders contribute to the ongoing evolution and optimization of livestock populations for various agricultural and food production systems.

Breeding Goals and Traits

Breeding goals and traits are fundamental considerations in animal breeding programs. These goals define the desired characteristics and attributes that breeders aim to improve through selective mating and genetic selection. The selection of breeding goals and traits depends on the specific objectives of the breeding program, market demands, and the intended use of the animals. Here are some common breeding goals and traits:

1. Productivity Traits:

- Growth Rate: Animals with higher growth rates reach market weight faster, improving efficiency.
- Milk Yield: Dairy animals with higher milk production contribute to increased dairy product output.

2. Carcass Traits:

- Meat Quality: Traits such as marbling, tenderness, and color affect the quality of meat for consumption.
- Carcass Yield: High carcass yield ensures efficient meat production from each animal.

3. Reproductive Traits:

- Fertility: Animals with higher fertility rates have shorter calving or farrowing intervals.
- Pregnancy Rate: Improved pregnancy rates result in shorter non-productive periods.

4. Disease Resistance and Health Traits:

- Resistance to Diseases: Animals with genetic

resistance to specific diseases require fewer medical interventions.

- Overall Health: Breeding for robust health and disease resistance minimizes production losses.

5. Environmental Adaptability:

- Heat Tolerance: Animals adapted to hot climates are better suited to thrive in such conditions.
- Cold Tolerance: Animals adapted to cold climates withstand lower temperatures.

6. Feed Efficiency:

- Feed Conversion Ratio: Animals with better feed conversion ratios convert feed into meat or milk more efficiently.
- Reduced Feed Requirements: Animals that require less feed for the same level of production contribute to resource savings.

7. Behavior and Temperament:

- Docility: Animals with calm and manageable temperaments are easier to handle and reduce stress during management practices.

8. Nutritional Traits:

- Nutrient Efficiency: Animals that utilize nutrients effectively contribute to better overall health and performance.
- Improved Nutrient Utilization: Efficient nutrient utilization reduces waste and promotes efficient growth.

9. Fiber Traits:

- Fiber Quality: In fiber-producing animals like sheep and goats, improved fiber quality enhances the value of the product.

10. Environmental Sustainability: - Reduced Environmental Impact: Breeding for reduced greenhouse gas emissions or resource-efficient traits contributes to sustainability.

11. Market Demand: - Consumer Preferences: Traits aligned with consumer preferences, such as lean meat or specific milk characteristics. - Value-Added Traits: Breeding for traits that add value to end products, like omega-3 fatty acids in eggs or milk.

12. Genetic Diversity: - Maintaining Genetic Diversity: Breeding goals may include maintaining genetic diversity to prevent inbreeding and maintain adaptability.

Selecting the right breeding goals and traits requires a thorough understanding of industry demands, consumer preferences, and the animals' natural behaviors. It also involves considering the genetic architecture of the traits, heritability, and the availability of genetic resources. Balanced breeding goals take into account multiple traits to achieve genetic improvement while avoiding unintended consequences. As breeding programs evolve, the definition of desirable traits may adapt to changing market trends, scientific advancements, and sustainability considerations.

Inbreeding and Crossbreeding

Inbreeding and crossbreeding are two contrasting breeding strategies used to influence the genetic composition of animal populations. Both approaches have specific advantages and challenges and are employed based on breeding goals, desired traits, and the overall objectives of the breeding program.

Inbreeding: Inbreeding involves mating animals that are closely related, such as siblings or cousins. While inbreeding can intensify the expression of both desirable and undesirable traits, it is often associated with a reduction in genetic diversity within a population.

Advantages of Inbreeding:

1. **Fixation of Desirable Traits:** Inbreeding can increase the chances of "fixing" desirable traits within a population, leading to greater uniformity.
2. **Testing of Traits:** Inbreeding can reveal recessive traits that may not be expressed in heterozygous individuals.
3. **Enhanced Homozygosity:** Inbreeding increases the likelihood of homozygosity, which can lead to uniformity and predictability of offspring traits.

Challenges of Inbreeding:

1. **Loss of Genetic Diversity:** Inbreeding reduces genetic diversity, increasing the risk of expressing recessive genetic disorders and susceptibility to diseases.
2. **Inbreeding Depression:** Cumulative inbreeding can lead to inbreeding depression, characterized

by reduced fitness, vitality, and reproductive performance.

3. **Undesirable Traits:** Inbreeding can intensify the expression of undesirable traits, including genetic defects and reduced performance.

Crossbreeding: Crossbreeding involves mating animals of different breeds or lines. The goal is to capitalize on the complementary strengths of different breeds, achieving hybrid vigor (heterosis) in the offspring.

Advantages of Crossbreeding:

1. **Hybrid Vigor:** Crossbred offspring often exhibit improved performance, health, and fertility due to hybrid vigor, resulting from the masking of detrimental recessive alleles.
2. **Complementary Traits:** Crossbreeding allows for the combination of desirable traits from different breeds to optimize productivity.
3. **Genetic Diversity:** Crossbreeding maintains or increases genetic diversity, reducing the risk of inbreeding-related issues.

Challenges of Crossbreeding:

1. **Heterogeneity:** Crossbred populations can exhibit greater variability in traits due to the mixture of different genetic backgrounds.
2. **Uniformity:** Crossbreeding can lead to reduced uniformity if not managed properly, as individuals from the same cross may vary.
3. **Complex Management:** Managing different breeds' specific needs and adapting to varying traits can be logistically challenging.

Considerations:

. The choice between inbreeding and crossbreeding

depends on breeding goals, the availability of genetic resources, and the need for genetic improvement.

- Inbreeding is suitable for consolidating specific traits within a breed but requires careful management to avoid negative consequences.
- Crossbreeding is effective for introducing new genetic material, enhancing performance, and managing genetic diversity.

In practice, breeders often employ a combination of inbreeding and crossbreeding strategies to achieve their goals while minimizing the potential drawbacks. The choice of strategy depends on the specific breeding program's objectives, available resources, and long-term breeding plan.

Reproduction and Reproductive Technologies

Reproduction and reproductive technologies play a crucial role in animal breeding and the management of livestock populations. These technologies enable breeders to control and enhance reproductive processes, optimize breeding outcomes, and achieve specific genetic goals. Here's an overview of reproduction and common reproductive technologies used in animal agriculture:

Natural Reproduction: Natural reproduction involves allowing animals to mate naturally without intervention. It relies on the natural behaviors and physiological processes of animals.

Artificial Insemination (AI): Artificial insemination involves the collection, processing, and deposition of semen from a male animal into a female's reproductive tract without natural mating. AI offers several benefits:

- Genetic Improvement: Allows breeders to use superior sires for multiple matings.
- Disease Control: Reduces the risk of disease transmission through natural mating.
- Record Keeping: Enables accurate record-keeping of mating dates and sire information.

Embryo Transfer (ET): Embryo transfer involves the collection of embryos from a genetically valuable female (donor) and their transfer to surrogate females (recipients) for gestation. ET has various applications:

- Rapid Genetic Progress: Allows donors to produce multiple offspring in a single year.
- Preservation of Genetics: Enables preservation of genetics from rare or valuable animals.
- Disease Control: Minimizes disease transmission compared to natural mating.

In Vitro Fertilization (IVF): IVF involves fertilizing oocytes (egg cells) outside the female's body and transferring the resulting embryos into surrogate recipients. IVF is used to:

- Multiply Genetic Material: Allows rapid multiplication of genetically valuable animals.
- Overcome Fertility Issues: Can bypass fertility issues in females.
- Genome Editing: Facilitates genome editing by manipulating embryos before transfer.

Semen Sexing: Semen sexing is a technology that separates sperm cells based on their sex chromosome content. This allows breeders to control the gender ratio of offspring, which can be beneficial for various production systems.

Cloning: Cloning involves creating genetically identical copies of an animal through somatic cell nuclear transfer. Cloning can be used to reproduce animals with exceptional genetics or performance traits.

Genomic Selection: Genomic selection uses genetic markers to predict an animal's breeding value for various traits. It allows breeders to select breeding animals at a young age, increasing the accuracy of selection.

Hormonal Manipulation: Hormonal manipulation involves using hormones to regulate reproductive processes, induce estrus (heat), synchronize estrus in groups of females, and improve conception rates.

Assisted Reproductive Technologies (ARTs): ARTs encompass various techniques that assist with reproduction, including cryopreservation (freezing) of semen and embryos, ovum pick-up (retrieval of oocytes), and in vitro embryo production.

Challenges and Considerations:

- Reproductive technologies require expertise, proper facilities, and adherence to ethical and animal welfare considerations.
- Implementation costs, technical expertise, and success rates vary depending on the technology.
- Breeding goals, species-specific traits, and market demands influence the choice of reproductive technologies.

Reproductive technologies continue to evolve, offering opportunities for breeders to make significant genetic advancements, improve breeding efficiency, and contribute to sustainable and productive livestock systems.

Natural Breeding Techniques

Natural breeding techniques involve allowing animals to mate naturally without human intervention. These techniques leverage the natural reproductive behaviors and physiological processes of animals to achieve successful breeding outcomes. While modern reproductive technologies offer advanced control over breeding, natural breeding remains a fundamental and cost-effective method used in animal agriculture. Here are some common natural breeding techniques:

1. Pasture Mating:

- Animals are allowed to graze together in open pastures or range areas.
- Mating occurs naturally as animals interact and form social groups.
- Common for cattle, sheep, and other grazing animals.

2. Pen Mating:

- Animals are placed in pens or enclosures where they can interact and mate.
- Used for various species, including swine, rabbits, and poultry.

3. Turnout Mating:

- Animals are temporarily separated before being released together to simulate a natural breeding scenario.
- Facilitates mating and reduces aggression or over-mating.

4. Free-Range Mating:

- Animals are given access to outdoor areas where they can mate and exhibit natural behaviors.
- Common for poultry, ducks, and other outdoor-reared animals.

5. Herd or Flock Breeding:

- Animals are managed in groups, and natural mating occurs within the group.
- Suitable for species that exhibit group mating behaviors, such as goats and some poultry.

6. Mating in Estrus:

- Female animals are observed for signs of estrus (heat), indicating they are ready for mating.
- Mating occurs when females exhibit receptive behaviors and physiological changes.

7. Bull or Ram Rotation:

- Multiple males are rotated among groups of females to ensure broader genetic diversity and maximize conception rates.

8. Natural Breeding Management:

- Monitoring: Regularly observe and monitor animals for signs of estrus or mating behavior.
- Record Keeping: Maintain accurate records of mating dates and parentage information.
- Separation: Separate males and females when mating is not desired to control breeding.

Advantages of Natural Breeding Techniques:

- Mimics Natural Behavior: Natural breeding techniques allow animals to engage in their natural reproductive behaviors.

- Low Cost: Compared to advanced reproductive technologies, natural breeding techniques are often more cost-effective.
- Minimizes Stress: Natural mating can be less stressful for animals compared to some assisted reproductive techniques.

Considerations:

- Some natural breeding techniques may lead to unwanted inbreeding if not managed properly.
- Depending on the species, gender ratios and behavioral interactions need careful management.
- Natural breeding may not always guarantee high rates of conception, and reproductive efficiency can be influenced by factors such as animal health and environment.

Natural breeding techniques remain a valuable approach for many livestock producers, allowing them to work with the animals' innate behaviors and natural reproductive processes while promoting genetic diversity and animal welfare.

Artificial Insemination

Artificial insemination (AI) is a reproductive technology that involves the collection, processing, and deposition of semen from a male animal into a female's reproductive tract to achieve pregnancy without natural mating. AI offers several benefits and is widely used in various animal species for genetic improvement, disease control, and efficient breeding management. Here's an overview of artificial insemination:

Advantages of Artificial Insemination:

1. **Genetic Improvement:**
   - AI enables the use of genetically superior sires to sire many offspring.
   - Breeders can access genetics from sires located remotely, leading to rapid genetic progress.
2. **Disease Control:**
   - AI reduces the risk of disease transmission compared to natural mating.
   - Quarantined and disease-free semen can be used to minimize disease spread.
3. **Record Keeping:**
   - Accurate record-keeping is facilitated, including mating dates, sire information, and production data.
4. **Economic Efficiency:**
   - AI allows breeders to optimize breeding resources, reduce labor, and minimize animal handling.
5. **Selective Breeding:**

- Breeders can select sires based on specific traits, such as growth rate, milk production, or disease resistance.

6. **Controlled Breeding Season:**
   - AI enables breeders to synchronize estrus and control breeding seasons, improving breeding efficiency.

Steps in the AI Process:

1. **Semen Collection:**
   - Semen is collected from a male animal (bull, boar, ram) using artificial vagina, electroejaculation, or other methods.

2. **Semen Processing:**
   - Semen is processed to remove debris, dilute the semen, and extend its shelf life.
   - Extenders help preserve sperm viability during storage and transportation.

3. **Semen Storage:**
   - Processed semen is stored in liquid nitrogen at very low temperatures to maintain sperm viability for extended periods.

4. **Estrus Detection:**
   - Female animals are monitored for signs of estrus (heat), such as increased activity and mounting behavior.

5. **Insemination:**
   - A trained technician deposits the processed semen into the female's reproductive tract using a catheter or AI gun.

6. **Pregnancy Diagnosis:**
   - Pregnancy is confirmed through methods such as rectal palpation, ultrasound, or hormone assays.

Considerations and Challenges:

1. **Semen Quality:** Semen quality, including sperm concentration and motility, impacts AI success.
2. **Timing:** Proper timing of AI is critical to ensure that semen is deposited during the female's fertile period.
3. **Skilled Personnel:** AI requires skilled technicians trained in semen handling, estrus detection, and insemination.
4. **Equipment and Facilities:** Proper equipment and facilities are essential for semen collection, processing, and AI procedures.
5. **Genetic Diversity:** While AI can enhance genetic improvement, maintaining genetic diversity is important to prevent inbreeding.

Artificial insemination is a valuable tool in animal breeding that offers breeders greater control over genetic selection, disease management, and reproductive efficiency. When applied correctly and with attention to detail, AI contributes to the genetic advancement of livestock populations and the achievement of breeding goals.

Embryo Transfer and Cloning

Embryo transfer (ET) and cloning are advanced reproductive technologies used in animal breeding to achieve specific genetic goals and enhance reproductive efficiency. These techniques offer the potential to rapidly propagate genetically valuable individuals and preserve rare or exceptional genetics. Here's an overview of embryo transfer and cloning:

Embryo Transfer (ET):

Embryo transfer involves the collection of embryos from a genetically valuable female (donor) and their subsequent transfer to surrogate females (recipients) for gestation and birth. ET allows breeders to multiply the genetic impact of superior females and accelerate genetic progress.

Steps in the Embryo Transfer Process:

1. **Superovulation of Donor Female:**
   - The donor female is treated with hormones to stimulate the development of multiple ovulatory follicles.
2. **Embryo Collection:**
   - The donor female's ovaries are surgically accessed, and embryos are aspirated or flushed from the reproductive tract.
3. **Embryo Evaluation and Processing:**
   - Embryos are evaluated for quality and stage of development.
   - Embryos are processed, sorted, and stored in preparation for transfer.
4. **Recipient Synchronization:**

- Surrogate females (recipients) are synchronized with the donor's estrus cycle.

5. **Embryo Transfer:**
   - Embryos are non-surgically transferred into the uterine horns of recipient females.
6. **Gestation and Birth:**
   - Recipient females carry the embryos to term and give birth to offspring.

Advantages of Embryo Transfer:

1. **Rapid Genetic Progress:**
   - High-quality genetics can be propagated quickly, leading to faster genetic improvement.
2. **Genetic Preservation:**
   - Rare or valuable genetics can be preserved by creating multiple embryos.
3. **Maximize Reproductive Potential:**
   - ET allows females to produce more offspring than naturally possible.
4. **Disease Control:**
   - Disease-free embryos can be transferred to surrogate females, reducing disease risk.

Challenges of Embryo Transfer:

1. **Technical Expertise:** ET requires skilled personnel for proper embryo handling, transfer, and synchronization.
2. **Synchronization:** Achieving proper synchronization between donors and recipients is crucial.
3. **Embryo Viability:** Not all embryos collected are viable or result in successful pregnancies.
4. **Cost:** ET can be expensive due to hormone treatments, labor, and facilities.

Cloning:

Cloning involves creating genetically identical copies of an animal through somatic cell nuclear transfer (SCNT). SCNT involves transferring the nucleus of a donor cell into an enucleated egg cell, which is then stimulated to develop into an embryo.

Advantages of Cloning:

1. **Genetic Replication:** Cloning produces genetically identical individuals with the same traits as the donor.
2. **Genetic Preservation:** Rare or exceptional genetics can be preserved through cloning.
3. **Research and Biotechnology:** Cloning has applications in research, pharmaceuticals, and biotechnology.

Challenges of Cloning:

1. **Low Efficiency:** Cloning success rates can be low due to developmental challenges and abnormalities.
2. **Genetic Diversity:** Cloning reduces genetic diversity, which can lead to susceptibility to diseases and environmental changes.
3. **Ethical and Welfare Concerns:** Cloning raises ethical considerations and animal welfare concerns.

Embryo transfer and cloning are advanced techniques that offer powerful tools for breeders to achieve specific breeding goals. While these technologies hold promise for genetic advancement and preservation, their application requires careful management, technical expertise, and ethical considerations.

Livestock Housing and Facilities

Livestock housing and facilities play a critical role in the management, health, and well-being of animals in animal agriculture. Proper housing and facilities provide a controlled environment that meets the animals' physiological, behavioral, and welfare needs, as well as the operational requirements of the farm. Here are key considerations and factors related to livestock housing and facilities:

1. Housing Types:

- Barns: Enclosed structures with controlled environments for protection against weather and predators.
- Shelters: Open or partially covered structures that offer shade and protection from the elements.
- Pastures: Open grazing areas that provide natural foraging and movement opportunities.

2. Species and Requirements:

- Different species have distinct housing requirements based on size, behavior, and physiology.
- Housing should cater to the animals' comfort, space, social interactions, and natural behaviors.

3. Housing Design:

- Proper ventilation, lighting, and temperature control are essential for animal health and productivity.
- Flooring materials should be chosen for hygiene, comfort, and prevention of injuries.

4. Space Allocation:

- Provide adequate space per animal to prevent overcrowding and facilitate movement.
- Space requirements vary by species, age, size, and intended use.

5. Comfort and Welfare:

- Housing should offer protection from extreme weather conditions, including cold, heat, wind, and rain.
- Bedding and resting areas should be clean, dry, and comfortable.

6. Ventilation:

- Proper ventilation ensures fresh air exchange, controls humidity, and minimizes respiratory issues.
- Natural ventilation, mechanical systems, and exhaust fans are used to manage air quality.

7. Lighting:

- Natural and artificial lighting influence animals' biological rhythms, behavior, and productivity.
- Adequate lighting enhances feeding, reproductive cycles, and overall well-being.

8. Feed and Water Management:

- Adequate feed and water availability is critical for animal health and performance.
- Automated feeding and watering systems optimize efficiency and reduce labor.

9. Waste Management:

- Proper waste management reduces odors, pests, and disease risks.

- Manure management and disposal strategies are essential for environmental sustainability.

10. Biosecurity and Disease Prevention: - Proper housing design and management practices help prevent the introduction and spread of diseases. - Quarantine areas and strict biosecurity measures safeguard animal health.

11. Handling Facilities: - Handling facilities, including chutes and pens, ensure safe and efficient animal handling for veterinary care, vaccination, and transportation.

12. Zoning and Regulations: - Compliance with local regulations and zoning laws ensures proper siting and construction of livestock facilities.

13. Expansion and Flexibility: - Design facilities with future growth and flexibility in mind to accommodate changes in herd size or management practices.

14. Sustainability: - Implement energy-efficient technologies, waste recycling, and resource conservation practices to reduce the environmental impact.

15. Animal Welfare: - Housing and facilities should prioritize the animals' well-being, addressing their behavioral, social, and physiological needs.

Properly designed and managed livestock housing and facilities contribute to animal health, production efficiency, and the overall sustainability of agricultural operations. Careful consideration of the animals' needs, environmental factors, and operational requirements ensures that livestock are provided with comfortable and appropriate living conditions.

Housing Designs for Different Species

Housing designs for different livestock species vary based on their specific physiological, behavioral, and environmental needs. Each species has unique requirements, and designing appropriate housing facilities is essential to ensure animal welfare, health, and productivity. Here are housing considerations for common livestock species:

1. Cattle:

- Barns or freestall systems for dairy cattle with comfortable resting areas, feed alleys, and milking parlors.
- Beef cattle may use open feedlots, pasture systems, or sheltered areas for protection against weather.
- Adequate ventilation to prevent respiratory issues and heat stress.

2. Pigs:

- Swine barns with proper ventilation, temperature control, and flooring that prevents injuries and provides comfort.
- Farrowing crates for sows and farrowing pens for piglets, promoting piglet safety and maternal care.
- Slatted flooring to allow waste removal, and appropriate enrichment to prevent boredom.

3. Poultry:

- Layer and broiler houses designed for optimized ventilation, lighting, and temperature control.

- Cages, aviaries, or free-range systems for egg-laying hens, allowing natural behaviors and nesting.
- Broiler houses with proper litter management, feeders, and waterers.

4. Sheep and Goats:

- Barns, sheds, or shelters that provide protection from weather and predators.
- Grazing systems with rotational pastures and shelter for lambing or kidding.
- Appropriate bedding and resting areas for comfort.

5. Horses:

- Stables or barns with well-ventilated stalls, feed storage, and tack rooms.
- Paddocks or pastures for exercise, social interaction, and grazing.
- Shelter for protection against extreme weather conditions.

6. Rabbits:

- Rabbit hutches or cages with proper ventilation and insulation.
- Individual housing to prevent aggression and ensure proper sanitation.
- Nest boxes for breeding and kindling.

7. Poultry (Chickens and Turkeys):

- Chicken coops or houses with nesting boxes and roosting perches.
- Turkeys may require larger facilities with proper ventilation and lighting.
- Brooder houses for young chicks with controlled temperature and proper bedding.

8. Bees:

- Hives or apiaries that protect bees from the weather and predators.
- Adequate ventilation and insulation to maintain hive health.
- Swarm-catching equipment and honey extraction facilities.

When designing livestock housing, it's important to consider the animals' natural behaviors, space requirements, social dynamics, reproductive needs, and health considerations. Proper ventilation, lighting, flooring, and waste management are critical components for all species. Additionally, housing should comply with animal welfare guidelines, biosecurity measures, and local regulations. Customized housing designs that cater to the specific needs of each species contribute to the well-being and success of livestock operations.

Environmental Considerations and Animal Welfare

Environmental considerations and animal welfare are closely interconnected in livestock production. Providing animals with appropriate housing, care, and management practices that align with their natural behaviors and physiological needs is not only essential for animal welfare but also has significant implications for environmental sustainability. Here's how environmental considerations and animal welfare are intertwined:

1. Housing and Space:

- Proper housing design that accommodates animals' space requirements promotes their physical and mental well-being.
- Sufficient space prevents overcrowding, stress, and aggression among animals.

2. Ventilation and Air Quality:

- Adequate ventilation ensures fresh air exchange and minimizes respiratory issues.
- Good air quality prevents the buildup of harmful gases and pathogens, promoting animal health.

3. Temperature Control:

- Proper temperature management in housing prevents heat stress or cold stress, which can compromise animal welfare and productivity.

4. Bedding and Flooring:

- Clean, comfortable bedding and appropriate flooring prevent injuries, lameness, and discomfort.
- Proper bedding and flooring also contribute to waste management and environmental hygiene.

5. Enrichment and Behavioral Needs:

- Providing enrichment items, such as scratching posts, toys, or structures, encourages natural behaviors and reduces boredom and stress.
- Meeting behavioral needs enhances animal welfare and can improve production outcomes.

6. Disease Prevention and Biosecurity:

- Good animal welfare practices, including biosecurity measures, help prevent disease outbreaks that can have environmental and economic impacts.

7. Grazing and Foraging Opportunities:

- Providing access to outdoor spaces, pastures, or forage areas allows animals to engage in natural behaviors like grazing and foraging.
- This contributes to their well-being and reduces the environmental impact of feed production.

8. Waste Management:

- Proper waste management prevents pollution and ensures that waste products do not negatively affect animal health or the environment.
- Responsible waste disposal practices contribute to sustainable livestock production.

9. Antibiotic and Medication Use:

- Implementing responsible use of antibiotics and medications in accordance with animal welfare principles helps prevent the development of antimicrobial resistance and safeguards animal health.

10. Ethical Considerations: - Ethical treatment of animals is closely tied to environmental considerations, as practices that prioritize animal welfare often align with sustainable and responsible resource use.

Promoting animal welfare not only benefits the animals themselves but also has broader implications for the environment and society. When animals are provided with appropriate care, housing, and management, they are more likely to thrive, exhibit natural behaviors, and produce efficiently. This can lead to reduced stress, improved health, and optimized production, contributing to the overall sustainability and ethical foundation of livestock agriculture.

Animal Health and Disease Management

Animal health and disease management are critical aspects of livestock production that ensure the well-being of animals, maintain productivity, and safeguard public health. Effective disease prevention, monitoring, and response strategies are essential for minimizing the impact of diseases on both animal welfare and the industry as a whole. Here are key considerations in animal health and disease management:

1. Disease Prevention:

- Biosecurity: Implement strict biosecurity measures to prevent the introduction and spread of diseases through animals, people, equipment, and vehicles.
- Vaccination: Develop and follow vaccination schedules to protect animals against common diseases.
- Quarantine: Isolate new animals before introducing them to the herd or flock to prevent disease transmission.
- Nutrition: Provide balanced diets to strengthen immune systems and reduce susceptibility to diseases.
- Stress Management: Minimize stress factors that can compromise immune function.

2. Disease Detection and Monitoring:

- Regular Health Checks: Conduct regular health checks to identify signs of illness, lameness, or other health issues.

- Diagnostic Testing: Use laboratory tests to diagnose diseases accurately and quickly.
- Record Keeping: Maintain detailed records of health status, treatments, vaccinations, and disease history.

3. Disease Management:

- Prompt Treatment: Administer appropriate treatments promptly under veterinary guidance.
- Isolation: Isolate sick animals to prevent the spread of disease within the herd or flock.
- Supportive Care: Provide supportive care, such as fluids and rest, to sick animals to aid recovery.
- Antimicrobial Stewardship: Use antibiotics judiciously and in accordance with veterinary recommendations to prevent antimicrobial resistance.

4. Zoonotic Disease Prevention:

- Some animal diseases can be transmitted to humans (zoonoses). Implement measures to prevent zoonotic disease transmission.
- Regular Handwashing: Practice proper hygiene, including handwashing, when handling animals.

5. Emergency Preparedness:

- Develop emergency response plans to address disease outbreaks or other health crises.
- Collaborate with veterinary authorities and other stakeholders to manage and contain disease outbreaks.

6. Surveillance and Reporting:

- Collaborate with veterinary authorities to participate in disease surveillance and reporting programs.
- Timely reporting of disease outbreaks helps control disease spread and protect the industry.

7. Education and Training:

- Train farm personnel in disease prevention, biosecurity practices, and proper animal handling.
- Stay informed about emerging diseases and best practices in disease management.

8. Veterinarian Involvement:

- Work closely with veterinarians for animal health consultations, disease diagnosis, and treatment recommendations.

9. Record Keeping and Traceability:

- Maintain accurate records of animal movements, treatments, and vaccinations to facilitate disease traceability.

Effective animal health and disease management practices are essential to ensure the long-term sustainability of livestock production. By prioritizing animal well-being, implementing preventive measures, and responding promptly to disease challenges, producers can minimize the impact of diseases on their operations and contribute to the health of the entire industry.

Preventive Measures and Vaccination

Preventive measures and vaccination are essential components of disease management in livestock production. These strategies help minimize the risk of disease outbreaks, reduce the impact of diseases on animal health and welfare, and safeguard the productivity and sustainability of the industry. Here's a closer look at preventive measures and vaccination in livestock:

Preventive Measures:

1. **Biosecurity:**
   - Implement biosecurity protocols to prevent the introduction and spread of diseases to and from the farm.
   - Restrict farm access, quarantine new animals, and control visitors' movement to limit disease transmission.
2. **Isolation and Quarantine:**
   - Isolate new or sick animals from the rest of the herd or flock to prevent disease spread.
   - Quarantine new animals before introducing them to the main group to monitor for signs of illness.
3. **Sanitation:**
   - Maintain clean and hygienic housing and facilities to reduce the presence of pathogens.
   - Properly manage waste and manure to prevent disease transmission.
4. **Nutrition and Management:**
   - Provide balanced nutrition to support strong immune systems and overall health.

- Avoid overstocking and implement proper animal management practices to minimize stress.

5. **Vector Control:**
   - Manage vectors such as insects and rodents that can carry and spread diseases.

Vaccination:

Vaccination involves the administration of vaccines to animals to stimulate their immune systems and provide protection against specific diseases. Vaccination is a proactive approach to disease prevention and is a common practice in livestock production.

Key Points about Vaccination:

1. **Vaccine Selection:**
   - Choose vaccines based on the prevalent diseases in the area, the specific species, and the age of the animals.
   - Consult with veterinarians to determine the most appropriate vaccines for your herd or flock.

2. **Vaccination Schedule:**
   - Follow a vaccination schedule recommended by veterinarians or animal health experts.
   - Administer booster shots as needed to maintain immunity.

3. **Proper Administration:**
   - Ensure vaccines are administered correctly, using proper injection techniques and equipment.
   - Follow storage and handling guidelines to maintain vaccine efficacy.

4. **Record Keeping:**
   - Maintain accurate records of vaccinations,

including the type of vaccine, date of administration, and batch number.

5. **Herd Immunity:**
   - Vaccinating a significant portion of the herd or flock can contribute to herd immunity, reducing disease spread.

6. **Regular Review:**
   - Regularly review and update vaccination protocols based on changes in disease risks and advancements in vaccine technology.

Preventive measures and vaccination work hand in hand to protect livestock from diseases. A comprehensive disease management strategy combines biosecurity practices, proper nutrition, vaccination, regular health checks, and collaboration with veterinarians. By taking a proactive approach to disease prevention, producers can enhance animal health, welfare, and productivity while maintaining the sustainability of their operations.

Diagnosis and Treatment of Common Diseases

Diagnosis and treatment of common diseases in livestock are critical for maintaining animal health and preventing economic losses in the livestock industry. Timely and accurate diagnosis allows for prompt treatment, minimizing the spread of diseases and their impact on both individual animals and the entire herd or flock. Here's an overview of the process of diagnosing and treating common diseases in livestock:

Diagnosis of Common Diseases:

1. **Clinical Signs:**
   - Observe animals for abnormal behaviors, symptoms, or physical changes that indicate illness.
   - Common clinical signs include fever, lethargy, decreased appetite, coughing, lameness, diarrhea, and changes in behavior.
2. **Physical Examination:**
   - Conduct a thorough physical examination to assess body condition, vital signs, and any visible abnormalities.
   - Palpate, auscultate, and observe animals closely for specific signs related to different organ systems.
3. **Laboratory Tests:**
   - Laboratory tests such as blood tests, fecal exams, and swab samples help diagnose

specific diseases, detect pathogens, and assess overall health.

- These tests provide valuable information about blood parameters, microbial presence, and other diagnostic indicators.

4. **Diagnostic Imaging:**
   - Radiography (X-rays), ultrasound, and other imaging techniques help visualize internal structures and identify abnormalities.
   - Imaging aids in diagnosing conditions such as fractures, organ enlargements, and obstructions.

5. **Necropsy (Post-Mortem Examination):**
   - When animals die unexpectedly or exhibit severe symptoms, a necropsy can provide valuable insights into the cause of death.
   - Post-mortem examinations help identify diseases and assess the overall health of the herd or flock.

Treatment of Common Diseases:

1. **Veterinary Consultation:**
   - Consult with a veterinarian to confirm the diagnosis and develop a treatment plan.
   - Veterinarians provide expertise in disease management and prescription of appropriate medications.

2. **Medication Administration:**
   - Administer medications, such as antibiotics, antiparasitics, anti-inflammatories, and antivirals, as prescribed by a veterinarian.
   - Follow dosage instructions and withdrawal periods to ensure proper treatment and avoid residues.

3. **Supportive Care:**

- Provide supportive care such as fluids, nutrition, and rest to aid in recovery.
- Supportive care can alleviate symptoms and help animals regain their strength.

4. Isolation:

- Isolate sick animals to prevent disease spread within the herd or flock.
- This reduces the risk of contagion and allows for closer monitoring and care.

5. Herd Management:

- Adjust management practices to prevent disease spread, such as improving hygiene and sanitation, adjusting housing conditions, and implementing biosecurity measures.

6. Follow-Up:

- Monitor the progress of treated animals and consult with a veterinarian to assess the effectiveness of the treatment plan.
- Adjust treatment if necessary based on animal response and veterinary guidance.

Effective diagnosis and treatment of common diseases in livestock require collaboration with veterinarians, adherence to recommended protocols, and attention to the health and welfare of the animals. By promptly addressing disease challenges and implementing proper management strategies, livestock producers can ensure the well-being of their animals and maintain the economic sustainability of their operations.

Grazing and Pasture Management

Grazing and pasture management are essential components of sustainable livestock production. Proper management of grazing areas and pastures ensures optimal nutrition, animal health, and environmental sustainability. Implementing effective grazing and pasture management practices can lead to improved animal welfare, increased productivity, and reduced environmental impact. Here are key considerations for grazing and pasture management:

1. Rotational Grazing:

- Divide pastures into smaller paddocks and rotate animals through them to prevent overgrazing and allow for grass regrowth.
- Rotational grazing maximizes forage utilization and minimizes soil compaction.

2. Stocking Rate:

- Determine the appropriate number of animals that can graze a pasture without causing overgrazing or degradation.
- Stocking rates vary based on factors such as pasture size, forage quality, and animal nutritional requirements.

3. Grazing Periods:

- Allow pastures to rest and recover between grazing periods to maintain healthy vegetation.
- Rest periods promote grass growth and prevent soil

erosion.

4. Forage Quality and Nutrition:

- Monitor forage quality to ensure animals receive adequate nutrition.
- Rotate animals to different pastures with varying forage species to provide diverse nutritional sources.

5. Weed and Pest Control:

- Implement strategies to control weeds, pests, and invasive plant species that can reduce pasture productivity.
- Integrated pest management practices minimize the need for chemical treatments.

6. Water Access:

- Ensure easy access to clean and fresh water for grazing animals.
- Strategically place water sources within pastures to prevent overgrazing around water points.

7. Fencing and Infrastructure:

- Install sturdy fencing to contain animals and protect pastures from overgrazing and trampling.
- Consider installing shade structures, windbreaks, and shelter to provide animals with protection from extreme weather.

8. Pasture Renovation:

- Periodically renovate pastures by overseeding with desired forage species and implementing soil improvement practices.
- Renovation improves forage quality and enhances pasture productivity.

9. Soil Health:

- Monitor soil health through soil testing and implement practices that improve soil structure, fertility, and water retention.
- Healthy soils support robust plant growth and contribute to healthy pastures.

10. Managed Grazing Plans: - Develop managed grazing plans that outline rotation schedules, stocking rates, and pasture improvement goals. - Managed grazing plans enhance pasture utilization and sustainability.

11. Monitoring and Adaptation: - Regularly assess pasture conditions, forage availability, and animal performance. - Adapt management practices based on changing conditions and seasonal variations.

12. Environmental Considerations: - Practice responsible grazing management to prevent soil erosion, protect water quality, and preserve natural habitats.

Effective grazing and pasture management require a combination of science, observation, and adaptation. By promoting sustainable grazing practices, livestock producers can optimize the use of available resources, improve animal health and welfare, and contribute to the overall environmental stewardship of their land.

Rotational Grazing Systems

Rotational grazing is a livestock management system that involves dividing pastures into smaller paddocks or grazing areas and rotating animals through these areas on a regular basis. This approach promotes efficient forage utilization, pasture health, and animal well-being while preventing overgrazing and soil degradation. Rotational grazing systems offer several benefits and can be adapted to various livestock species and land types. Here are the key features and benefits of rotational grazing systems:

Key Features of Rotational Grazing Systems:

1. **Paddock Division:** Pastures are divided into smaller paddocks or grazing areas using fencing or temporary barriers.
2. **Rotation Schedule:** Animals are moved from one paddock to another in a planned sequence based on a rotation schedule.
3. **Rest Periods:** After grazing a paddock, it is allowed to rest and recover before animals return for another grazing cycle.
4. **Forage Regrowth:** Rest periods allow grasses and forage plants to regrow, maintaining adequate forage quality and quantity.
5. **Reduced Overgrazing:** Animals graze only a portion of the pasture, preventing overgrazing that can damage plants and soil.
6. **Improved Nutrient Distribution:** Rotational grazing helps distribute animal waste more evenly across the pasture, benefiting soil fertility.

Benefits of Rotational Grazing Systems:

1. **Optimal Forage Utilization:** Rotational grazing maximizes the use of available forage by preventing selective grazing and underutilization of certain areas.
2. **Enhanced Forage Quality:** Rest periods promote the growth of nutritious, tender forage that meets animals' nutritional needs.
3. **Animal Health and Welfare:** Animals have access to high-quality forage, clean water, and reduced parasite exposure, leading to improved health and well-being.
4. **Soil Health:** Controlled grazing reduces soil compaction and erosion, enhancing soil structure and promoting water infiltration.
5. **Sustainable Pasture Management:** Rotational grazing extends the life of pastures by preventing degradation and promoting healthy plant growth.
6. **Improved Productivity:** Animals gain weight more efficiently when provided with high-quality forage, leading to increased productivity.
7. **Reduced Input Costs:** Improved forage utilization reduces the need for supplemental feeding and reduces costs associated with pasture management.
8. **Environmental Benefits:** Rotational grazing systems contribute to environmental sustainability by minimizing soil erosion, improving water quality, and preserving biodiversity.
9. **Flexibility:** Rotational grazing systems can be adapted to different livestock species, pasture types, and land sizes.

Types of Rotational Grazing Systems:

1. **Time-Controlled Grazing:** Animals are rotated based on a fixed schedule, and paddocks are divided into equal time intervals.

2. **Stocking Density Rotations:** Paddock rotation is determined by the number of animals present rather than a fixed schedule.
3. **Mob Grazing:** Animals are moved frequently and densely graze small areas, followed by a longer rest period.
4. **High-Intensity, Low-Frequency Grazing:** Animals graze intensely for a short period, followed by a longer rest period.
5. **Adaptive Grazing Management:** Grazing intensity and rest periods are adjusted based on forage growth, weather conditions, and other factors.

Rotational grazing systems require careful planning, monitoring, and adjustment to ensure effective implementation. By optimizing forage utilization, promoting pasture health, and enhancing animal welfare, rotational grazing contributes to sustainable and profitable livestock production.

Forage Crops and Pasture Health

Forage crops and pasture health play a crucial role in providing livestock with nutritious feed, supporting animal well-being, and ensuring the sustainability of livestock production. Proper management of forage crops and pastures involves selecting appropriate plant species, optimizing growth conditions, and implementing strategies to maintain the health and productivity of grazing areas. Here are key considerations for forage crops and pasture health:

Forage Crop Selection:

1. **Species and Varieties:** Choose forage species and varieties that are well-suited to the local climate, soil type, and intended livestock species.
2. **Nutritional Content:** Select forage crops with high nutritional content, including protein, fiber, energy, and essential nutrients.
3. **Diversity:** Incorporate a mix of forage species to provide a balanced diet and reduce the risk of pest and disease issues.
4. **Cool-Season vs. Warm-Season:** Consider cool-season grasses (e.g., fescue, ryegrass) for spring and fall grazing and warm-season grasses (e.g., Bermuda grass, bahiagrass) for summer grazing.

Pasture Establishment and Maintenance:

1. **Soil Preparation:** Prepare the soil by testing its pH, fertility, and drainage to ensure optimal forage growth.

2. **Seeding:** Use appropriate seeding methods, such as broadcasting or drilling, to establish forage crops.

3. **Fertilization:** Apply fertilizers based on soil test results to provide essential nutrients for healthy plant growth.

4. **Weed Management:** Implement weed control strategies to prevent weed competition and ensure forage quality.

5. **Pest Management:** Monitor for pests such as insects and rodents that can damage forage crops.

6. **Grazing Management:** Rotate animals through pastures to prevent overgrazing, allow forage regrowth, and maintain pasture health.

Pasture Health and Maintenance:

1. **Monitoring:** Regularly assess pasture conditions, forage availability, and plant health.

2. **Rest Periods:** Provide adequate rest periods to allow forage plants to recover and regrow.

3. **Overgrazing Prevention:** Avoid overgrazing, which can lead to soil compaction, reduced plant vigor, and increased weed pressure.

4. **Soil Health:** Practice soil conservation techniques to maintain soil structure, fertility, and water retention.

5. **Irrigation:** Provide supplemental irrigation during dry periods to prevent drought stress and promote forage growth.

6. **Reseeding and Renovation:** Reseed areas with thinning vegetation to maintain pasture productivity.

7. **Grazing Height:** Maintain a minimum grazing height to avoid weakening plants and allow for quick regrowth.

8. **Biodiversity:** Encourage the growth of legumes (e.g., clover, alfalfa) alongside grasses to enhance soil nitrogen levels and improve forage quality.

9. **Manure Management:** Manage animal waste to prevent overloading pastures with nutrients and maintain soil health.
10. **Soil Erosion Control:** Implement erosion control measures, such as contouring and cover cropping, to prevent soil erosion.

Healthy forage crops and pastures contribute to improved animal nutrition, reduced feed costs, and sustainable livestock production. By focusing on soil health, proper grazing management, and regular monitoring, producers can ensure that their pastures provide high-quality forage for livestock and maintain the long-term health of the land.

Dairy Cattle Management

Dairy cattle management involves a comprehensive approach to caring for dairy cows to ensure their health, productivity, and well-being while optimizing milk production. Effective dairy cattle management encompasses various aspects, including nutrition, housing, health care, reproduction, and overall herd management. Here are key considerations for successful dairy cattle management:

1. Nutrition:

- Provide balanced diets that meet the nutritional requirements of different stages of lactation, growth, and reproduction.
- Incorporate high-quality forage, grains, protein sources, vitamins, and minerals into the diet.
- Work with nutritionists to formulate rations that optimize milk production while maintaining cow health.

2. Housing and Facilities:

- Provide clean and comfortable housing that protects cows from adverse weather conditions.
- Offer adequate ventilation and proper temperature control to prevent heat stress.
- Design housing layouts that allow for easy access to feed, water, and resting areas.

3. Health Care:

- Implement a comprehensive health care program that

includes vaccination, parasite control, and disease prevention.
- Conduct regular health checks and monitor for signs of illness or distress.
- Collaborate with veterinarians to establish health protocols and manage diseases effectively.

4. Reproduction:

- Implement an efficient reproductive program to maximize conception rates and calving intervals.
- Monitor heat cycles, use artificial insemination, and employ advanced reproductive technologies as needed.
- Manage pregnancy and calving to ensure the health of both cows and newborn calves.

5. Milking Management:

- Establish consistent milking routines and practices to prevent stress and ensure milk quality.
- Maintain clean and sanitized milking equipment to prevent mastitis and maintain milk hygiene.
- Monitor milk production and quality, and promptly address any deviations.

6. Herd Management:

- Keep detailed records of individual cow health, reproduction, and production data.
- Implement a record-keeping system to track cow performance and make informed management decisions.
- Use herd management software to analyze data and identify trends for improvement.

7. Transition Management:

- Pay special attention to cows during the transition

period from dry period to early lactation.

- Manage nutrition, health, and comfort to prevent metabolic disorders and ensure successful calving.

8. Calf Management:

- Provide proper care for newborn calves, including colostrum feeding, proper housing, and health management.
- Implement calf health protocols to minimize mortality and promote growth.

9. Environmental Considerations:

- Implement sustainable practices to minimize the environmental impact of dairy operations.
- Manage waste, nutrient runoff, and effluents to protect water quality and ecosystems.

10. Continuous Learning and Improvement: - Stay updated with the latest advancements in dairy cattle management practices. - Attend workshops, conferences, and training programs to enhance skills and knowledge.

Effective dairy cattle management requires a combination of expertise, attention to detail, and dedication to the well-being of the cows. By providing optimal nutrition, health care, and reproduction management, dairy producers can ensure that their cattle are healthy, productive, and contribute to the success of the dairy operation.

Milk Production and Milking Techniques

Milk production and milking techniques are crucial components of dairy cattle management. Efficient milking practices ensure the well-being of cows, maintain milk quality, and optimize milk production. Proper milking techniques, equipment maintenance, and hygiene protocols contribute to the overall success of a dairy operation. Here are key considerations for milk production and milking techniques:

1. Milking Equipment:

- Ensure milking equipment, including milking machines, pulsators, and udder cups, is properly maintained and sanitized.
- Regularly check for proper vacuum levels, pulsation rates, and functioning of equipment components.

2. Milking Routine:

- Establish a consistent milking routine that minimizes stress and discomfort for cows.
- Follow a standardized process for pre-milking preparation, milking, and post-milking procedures.

3. Pre-Milking Preparation:

- Properly clean and sanitize udders before milking to prevent contamination and maintain milk quality.
- Use a pre-milking teat disinfectant to reduce the risk of mastitis-causing pathogens.

4. Milking Techniques:

- Use gentle and consistent hand or machine milking techniques to avoid discomfort and potential injury to cows.
- Pay attention to udder health and milk flow rates during milking.

5. Milking Frequency:

- Determine the appropriate milking frequency based on the lactation stage of cows and production goals.
- Most dairy cows are milked two to three times daily.

6. Milk Quality:

- Maintain strict hygiene to prevent contamination and ensure high milk quality.
- Monitor somatic cell counts (SCC) and bacterial levels to assess udder health and milk hygiene.

7. Teat Health:

- Monitor teat health and condition to prevent injuries and mastitis.
- Use teat dip solutions to maintain teat skin health and prevent bacterial infections.

8. Milker Hygiene:

- Milkers should practice proper hand hygiene, wear clean clothing, and adhere to hygiene protocols.
- Prevent cross-contamination between cows by changing gloves and sanitizing between animals.

9. Milk Storage and Cooling:

- Promptly transfer milk to a clean, well-maintained bulk tank for cooling and storage.
- Maintain proper cooling temperatures to preserve milk

quality.

10. Training and Supervision: - Train milkers in proper milking techniques, equipment operation, and hygiene practices. - Provide ongoing supervision and feedback to ensure consistent milking procedures.

11. Record Keeping: - Keep detailed records of individual cow milk production, milk quality, and udder health. - Use data to identify trends, make informed decisions, and address issues promptly.

12. Employee Welfare: - Provide training, support, and ergonomic milking equipment to ensure milker well-being.

Efficient milking techniques not only contribute to milk production and quality but also enhance the comfort and well-being of dairy cows. By adhering to best practices, investing in equipment maintenance, and maintaining high standards of hygiene, dairy producers can achieve optimal milk production while promoting animal health and welfare.

Dairy Cow Nutrition and Health

Dairy cow nutrition and health are intertwined aspects of successful dairy cattle management. Providing balanced and appropriate nutrition plays a crucial role in maintaining the health, well-being, and productivity of dairy cows. A well-designed nutrition program, combined with effective health management practices, ensures optimal milk production, reproductive performance, and overall herd health. Here's a closer look at dairy cow nutrition and health considerations:

Dairy Cow Nutrition:

1. **Balanced Diet:**
   - Develop and provide balanced diets that meet the nutritional needs of different lactation stages, growth, and reproduction.
   - Ensure diets contain the right proportions of energy, protein, fiber, vitamins, and minerals.

2. **Forage Quality:**
   - Incorporate high-quality forage, such as hay and silage, into the diet to support rumen health and overall nutrition.
   - Monitor forage quality and adjust rations as needed.

3. **Concentrates and Grains:**
   - Include grains and concentrates to provide energy and essential nutrients for milk production.
   - Adjust the amount and composition of concentrates based on cow performance and lactation stage.

4. **Protein Sources:**
   - Use high-quality protein sources like soybean meal and canola meal to meet cows' protein requirements.
   - Balance amino acids to optimize milk production and maintain body condition.

5. **Minerals and Vitamins:**
   - Ensure adequate intake of essential minerals and vitamins through balanced rations and mineral supplementation.

6. **Water Access:**
   - Provide clean and fresh water at all times to support milk production, digestion, and overall cow health.

Dairy Cow Health:

1. **Routine Health Checks:**
   - Perform regular health checks to identify early signs of disease, lameness, or discomfort.
   - Monitor body condition scores and assess overall cow well-being.

2. **Vaccination and Disease Prevention:**
   - Implement a vaccination program to protect against common diseases and prevent disease outbreaks.
   - Follow recommended vaccination schedules and consult with veterinarians.

3. **Parasite Control:**
   - Develop and implement parasite control measures to minimize internal and external parasites.
   - Rotate deworming medications to prevent parasite resistance.

4. **Mastitis Prevention:**

- Practice proper udder hygiene and teat preparation during milking to prevent mastitis.
- Monitor somatic cell counts (SCC) and address high counts promptly.

5. **Reproductive Health:**
   - Maintain reproductive health through proper heat detection, artificial insemination, and monitoring of estrus cycles.
   - Address reproductive issues promptly to ensure successful calving intervals.

6. **Lameness Management:**
   - Implement hoof care practices to prevent and manage lameness.
   - Maintain clean and dry walking areas to prevent hoof issues.

7. **Transition Cow Management:**
   - Pay special attention to cows during the transition period from dry period to lactation.
   - Manage nutrition, health, and comfort to prevent metabolic disorders.

8. **Emergency Preparedness:**
   - Develop emergency response plans for disease outbreaks or other health emergencies.

Dairy cow nutrition and health management require collaboration between nutritionists, veterinarians, and dairy managers. By providing proper nutrition, preventive care, and prompt treatment, dairy producers can ensure the overall health, productivity, and welfare of their cows, leading to successful and sustainable dairy operations.

Dairy Products and Processing

Dairy products and processing involve transforming raw milk into a wide range of consumable dairy foods. Dairy processing ensures product safety, quality, and shelf life while creating value-added products that meet consumer preferences. From milk to various dairy products, such as cheese, yogurt, butter, and more, processing techniques play a crucial role in delivering nutritious and palatable dairy foods to consumers. Here's an overview of dairy products and the processing methods involved:

1. Pasteurization:

- Pasteurization is a heat treatment process that destroys harmful microorganisms while preserving the quality of milk.
- It ensures the safety of dairy products by reducing the risk of pathogens and extending shelf life.

2. Homogenization:

- Homogenization breaks down fat globules in milk into smaller sizes, preventing cream separation.
- It creates a uniform texture and consistency in dairy products like milk and ice cream.

3. Milk Products:

- Whole Milk: Fresh milk with natural fat content.
- Skim Milk: Milk with fat removed, suitable for low-fat diets.
- Low-Fat Milk: Milk with reduced fat content.

- Reduced-Lactose Milk: Milk with lowered lactose content for lactose-intolerant individuals.

4. Dairy Product Processing:

a. **Cheese Production:** - Cheese is made by coagulating milk proteins using enzymes or acid, followed by separating the curd from the whey. - Various types of cheeses, such as cheddar, mozzarella, and Swiss, undergo different aging and curing processes.

b. **Yogurt Production:** - Yogurt is produced by fermenting milk with specific bacterial cultures that convert lactose into lactic acid. - Flavors, fruits, and sweeteners can be added to enhance taste.

c. **Butter Production:** - Butter is made by churning cream to separate the fat from the liquid portion. - Different types of butter, including salted and unsalted, are available.

d. **Ice Cream Production:** - Ice cream is made by combining milk, cream, sugar, and flavorings, then freezing while being churned. - Various flavors, textures, and mix-ins create a wide range of ice cream options.

e. **Condensed and Evaporated Milk:** - These products are made by removing water from milk to create a concentrated product. - Condensed milk contains added sugar, while evaporated milk does not.

f. **Dairy Desserts and Cultured Products:** - Puddings, custards, and flans are dairy-based desserts made by thickening milk or cream with starch or eggs. - Cultured dairy products include sour cream, crème fraîche, and kefir.

5. Product Packaging:

- Dairy products are packaged in various formats, including cartons, bottles, cups, tubs, and more.
- Packaging materials should be safe, durable, and

capable of maintaining product quality.

6. Quality Control:

- Dairy processing plants implement quality control measures to ensure products meet regulatory standards and consumer expectations.
- Regular testing and inspection are conducted to monitor product quality, safety, and consistency.

7. Distribution and Storage:

- Dairy products are distributed to retail stores, supermarkets, and other outlets.
- Proper storage conditions, including temperature control, are essential to maintain product freshness.

Dairy processing is a dynamic and sophisticated field that involves various techniques and technologies to create a wide range of dairy products enjoyed by consumers worldwide. The safety, quality, and nutritional attributes of these products rely on well-managed processing practices and adherence to industry standards.

Pasteurization and Homogenization

Pasteurization and homogenization are two important processes in dairy processing that help ensure the safety, quality, and shelf life of dairy products while improving their texture and consistency. These processes are commonly applied to milk and other dairy products to address microbial contamination, separate fat globules, and enhance product attributes. Here's a closer look at pasteurization and homogenization:

Pasteurization: Pasteurization is a heat treatment process used to destroy harmful microorganisms present in raw milk while preserving its nutritional value and sensory properties. The primary objective of pasteurization is to ensure the safety of dairy products by reducing the risk of foodborne pathogens. The process involves heating the milk to a specific temperature for a set duration and then rapidly cooling it. There are two main methods of pasteurization:

1. **High-Temperature Short-Time (HTST) Pasteurization:**
   - In HTST pasteurization, milk is heated to around 161°F (72°C) for 15-20 seconds, followed by rapid cooling.
   - This method effectively kills harmful bacteria while retaining the milk's quality attributes.
2. **Ultra-High-Temperature (UHT) Pasteurization:**
   - UHT pasteurization involves heating milk to temperatures around 280-302°F (138-150°C) for a few seconds.
   - This process extends shelf life significantly and eliminates the need for refrigeration

until the package is opened.

Pasteurization eliminates or reduces pathogenic bacteria and enzymes that can spoil milk and affect its quality. It allows milk to be stored longer without refrigeration while still being safe for consumption.

Homogenization: Homogenization is a mechanical process that breaks down fat globules in milk into smaller sizes, preventing cream separation. Milk fat naturally tends to rise to the surface over time, leading to a non-uniform appearance in milk products. Homogenization ensures a uniform distribution of fat throughout the milk, resulting in a consistent texture and appearance. The process involves forcing milk through a narrow nozzle at high pressure, which breaks down fat globules without affecting other milk components.

Benefits of homogenization include:

- Improved Texture: Homogenized milk is smoother and creamier due to the smaller fat globule size.
- Enhanced Stability: Homogenization prevents cream separation, allowing milk to remain uniform over time.
- Better Whipping Properties: Cream from homogenized milk can be more easily whipped for desserts.
- Enhanced Taste and Mouthfeel: Homogenization can improve the sensory experience of milk products.

Both pasteurization and homogenization are essential steps in dairy processing, contributing to the safety, quality, and consumer satisfaction of a wide range of dairy products, including milk, yogurt, ice cream, and more. The choice of processing methods depends on the type of dairy product and its intended use.

Cheese, Butter, and Yogurt Production

Cheese, butter, and yogurt are popular dairy products that undergo specific production processes to transform raw milk into delicious and nutritious foods. Each of these products involves unique techniques that contribute to their distinct flavors, textures, and characteristics. Here's an overview of the production processes for cheese, butter, and yogurt:

Cheese Production:

Cheese production involves the coagulation of milk proteins, which leads to the formation of curds and whey. The curds are separated from the whey and then processed to create various types of cheese with different flavors, textures, and aging profiles. Here's a general outline of the cheese-making process:

1. **Coagulation:** Milk is heated and enzymes or acid are added to coagulate the milk proteins. The coagulated milk forms curds and whey.
2. **Cutting and Draining:** The curds are cut into smaller pieces to release whey and create a desired curd size. The whey is drained off.
3. **Heating and Cooking:** The curds are heated and stirred to expel additional whey and reach the desired moisture content.
4. **Salting:** Salt is added to the curds for flavor and preservation.
5. **Molding and Pressing:** Curds are placed in molds to give the cheese its shape and pressed to remove excess whey.
6. **Aging:** Cheese is aged for various periods, during

which flavors and textures develop.

7. **Ripening:** During aging, beneficial bacteria and molds interact with the cheese, contributing to its flavor profile.

Cheese varieties are created through variations in milk type, coagulants, processing techniques, and aging conditions.

Butter Production:

Butter is made by separating milk fat from the liquid portion of cream. The process involves churning the cream until the fat globules cluster together and form butter. Here's an outline of butter production:

1. **Separation:** Cream is separated from whole milk using centrifugation or gravity-based methods.
2. **Ripening:** The cream is allowed to ripen to develop the flavor of the butter.
3. **Churning:** The cream is agitated vigorously until the fat globules clump together and form butter. The liquid portion left behind is buttermilk.
4. **Washing:** The newly formed butter is washed with cold water to remove excess buttermilk and improve shelf life.
5. **Working:** The butter is kneaded to remove remaining moisture and create a uniform texture.

Butter can be produced in its natural form or modified by adding salt or other flavorings.

Yogurt Production:

Yogurt is made through the fermentation of milk by specific bacterial cultures that convert lactose into lactic acid. The fermentation process gives yogurt its tangy flavor and characteristic texture. Here's a simplified overview of yogurt production:

1. **Heating:** Milk is heated to a specific temperature to denature proteins and promote thickening.
2. **Cooling:** The milk is cooled to a temperature suitable for adding yogurt cultures.
3. **Inoculation:** Starter cultures containing specific bacterial strains are added to the milk to initiate fermentation.
4. **Fermentation:** The bacteria convert lactose into lactic acid, causing the milk to thicken and develop its characteristic flavor.
5. **Cooling and Packaging:** The yogurt is cooled, and additional ingredients such as fruit, flavorings, or sweeteners are added before packaging.

The choice of starter cultures and fermentation conditions can influence the flavor, texture, and probiotic content of yogurt.

Each of these dairy products requires specific processing techniques and careful attention to quality and safety standards. The diverse flavors and textures created by these processes contribute to the rich variety of dairy products available to consumers.

Meat Animal Production

Meat animal production involves the breeding, raising, and management of animals specifically for the purpose of producing meat for human consumption. It encompasses various livestock species, including cattle, pigs (swine), sheep, goats, and poultry, each of which has its unique management practices. Meat animal production aims to efficiently convert feed resources into high-quality meat products while ensuring animal welfare, health, and sustainable practices. Here's an overview of meat animal production:

1. Cattle Production:

- Beef cattle are raised primarily for their meat (beef).
- Calves are born, raised, and managed through different production stages, including cow-calf, stocker, and feedlot operations.
- Breeding focuses on improving genetics for meat quality, growth rate, and feed efficiency.

2. Swine (Pig) Production:

- Pigs are raised for pork, which is a widely consumed meat product.
- Swine production involves farrowing (birthing), nursery, and finishing stages.
- Modern swine production emphasizes controlled environments, nutrition, and disease management.

3. Sheep and Goat Production:

- Sheep and goats are raised for lamb, mutton, and goat

meat (chevon or goat meat).
- Grazing and browsing animals, they can utilize varied forage resources.
- Meat from sheep and goats is popular in many cultures and cuisines.

4. Poultry Production:

- Poultry includes chickens, turkeys, ducks, and other birds raised for meat.
- Broiler chickens are raised for meat, while layers are raised for egg production.
- Modern poultry production often involves controlled environments, specialized diets, and efficient growth rates.

Key Considerations in Meat Animal Production:

1. **Genetics and Breeding:**
   - Selecting and breeding animals with desirable traits such as growth rate, feed efficiency, meat quality, and disease resistance.
   - Genetic improvement programs are used to optimize production traits.
2. **Feeding and Nutrition:**
   - Providing balanced diets that meet the nutritional requirements of animals at different life stages.
   - Formulating diets to optimize growth, muscle development, and overall health.
3. **Housing and Management:**
   - Providing suitable housing and management systems that ensure animal comfort, health, and safety.
   - Proper ventilation, temperature control, and sanitation are crucial.

4. **Health and Disease Management:**
   - Implementing vaccination, biosecurity, and disease prevention programs to maintain animal health.
   - Addressing disease outbreaks promptly to minimize impact.

5. **Welfare Considerations:**
   - Ensuring animals are raised in humane conditions that meet their physiological and behavioral needs.
   - Addressing animal welfare concerns, including space requirements, enrichment, and handling.

6. **Sustainability and Environmental Impact:**
   - Implementing practices that minimize environmental impact, such as efficient resource utilization and waste management.
   - Sustainable practices contribute to responsible meat production.

7. **Quality Assurance and Food Safety:**
   - Following quality assurance standards to ensure the safety, quality, and traceability of meat products.
   - Maintaining hygiene and food safety practices during processing and distribution.

Meat animal production plays a significant role in providing a valuable protein source for human consumption. Responsible and efficient management practices are essential to meet the growing demand for meat products while considering animal welfare, environmental impact, and consumer preferences.

Raising Cattle, Sheep, Goats, and Pigs for Meat

Raising cattle, sheep, goats, and pigs for meat involves specific management practices tailored to each species' unique characteristics and requirements. Each species has distinct dietary preferences, growth rates, and environmental needs. Proper management ensures optimal growth, health, and meat quality. Here's an overview of raising these animals for meat:

1. Cattle:

- Cattle are raised primarily for beef production.
- Breeding: Select beef cattle breeds based on growth rate, meat quality, and adaptability to the local environment.
- Grazing: Cattle can graze on pastures, and rotational grazing systems optimize forage utilization.
- Feed: Provide a balanced diet with appropriate energy and protein levels. In feedlots, grain-based diets are common to promote rapid growth.
- Housing: Depending on the region, cattle may be raised on pasture or in feedlots with proper shelter.

2. Sheep:

- Sheep are raised for lamb (young sheep) and mutton (mature sheep) meat.
- Breeding: Choose sheep breeds suitable for meat production and adaptability to local conditions.
- Grazing: Sheep are efficient grazers and can utilize a

variety of forage resources.

- Feed: Supplement grazing with well-balanced diets for optimal growth and meat quality.
- Lambing: Provide shelter and manage lambing to ensure newborn lambs' health and survival.

3. Goats:

- Goats are raised for goat meat (chevon) and are known for their adaptability to diverse environments.
- Breeding: Select meat goat breeds for growth rate, meat quality, and disease resistance.
- Grazing and Browsing: Goats are browsers and can thrive on shrubs, weeds, and other vegetation.
- Feed: Supplement browse with balanced diets to meet nutritional requirements.
- Parasite Management: Goats are susceptible to internal parasites, so proper management is essential.

4. Pigs (Swine):

- Pigs are raised for pork production and are known for their efficient growth rates.
- Breeding: Choose pig breeds with good meat quality and growth characteristics.
- Housing: Pigs can be raised in various housing systems, including outdoor pens and indoor facilities.
- Feed: Provide balanced diets that promote rapid growth and meat quality.
- Disease Management: Implement biosecurity measures to prevent disease outbreaks.
- Reproduction: Manage breeding and farrowing for efficient piglet production.

Key Considerations for Meat Animal Production:

1. **Health and Welfare:**
   - Maintain animal health through vaccination,

disease prevention, and proper veterinary care.
- Provide appropriate housing, ventilation, and environmental conditions that promote animal welfare.

2. **Nutrition:**
   - Develop balanced diets that meet each species' nutritional requirements and support growth and meat quality.

3. **Genetics:**
   - Select breeds and genetics that align with meat production goals, adaptability, and local conditions.

4. **Reproduction:**
   - Implement effective breeding programs to optimize reproduction rates and ensure consistent meat supply.

5. **Handling and Management:**
   - Train handlers in proper animal handling techniques to minimize stress and ensure animal safety.

6. **Environmental Impact:**
   - Implement sustainable practices that minimize environmental impact, such as waste management and land use.

Raising cattle, sheep, goats, and pigs for meat requires knowledge, experience, and a commitment to animal welfare, food safety, and sustainable practices. Careful management and attention to each species' needs contribute to producing high-quality meat products that meet consumer demand.

Factors Influencing Meat Quality

Meat quality is influenced by a combination of factors that encompass various aspects of animal production, handling, processing, and even genetics. These factors collectively determine the sensory, nutritional, and safety attributes of meat products. Here are some key factors that influence meat quality:

1. Genetics:

- The genetic makeup of the animal plays a significant role in determining meat quality traits such as tenderness, flavor, and marbling.
- Breeding programs and selection for desirable traits can impact meat quality characteristics.

2. Animal Age and Species:

- Meat from younger animals often has more tender and less tough muscle fibers.
- Different species have distinct meat characteristics, flavor profiles, and textures.

3. Nutrition:

- A balanced and nutritious diet is essential for healthy animal growth and optimal meat quality.
- Proper nutrition contributes to desirable meat color, marbling, and flavor.

4. Pre-Slaughter Stress and Handling:

- Stress and rough handling before slaughter can lead to elevated levels of stress hormones, which may affect meat quality.

- Minimizing stress during transportation and handling helps maintain meat quality.

5. Slaughter Practices:

- Proper and humane slaughter practices are crucial for minimizing stress and ensuring meat quality.
- Fast and efficient slaughter methods help preserve meat freshness.

6. Aging and Maturation:

- Aging meat after slaughter allows natural enzymes to break down muscle fibers, enhancing tenderness and flavor.
- Controlled aging periods for specific meat cuts can optimize meat quality.

7. Marbling and Fat Content:

- Intramuscular fat (marbling) contributes to meat tenderness, juiciness, and flavor.
- Higher marbling levels are often associated with premium cuts of meat.

8. pH Levels:

- Post-slaughter pH levels affect meat texture, color, and water-holding capacity.
- Rapid pH decline after slaughter can lead to tougher meat.

9. Processing and Cooking Methods:

- Proper processing and cooking techniques can impact meat tenderness, juiciness, and flavor.
- Overcooking can result in dry and less flavorful meat.

10. Handling and Storage: - Proper handling and storage

practices are essential to prevent contamination and maintain meat freshness. - Temperature control is critical to prevent bacterial growth and spoilage.

11. Packaging: - Packaging materials and methods influence meat preservation, color retention, and protection from oxygen and moisture.

12. Certification and Traceability: - Meat quality assurance programs and traceability systems help ensure safe and high-quality meat products. - Certified meats meet specific quality and safety standards.

13. Consumer Preferences: - Consumer preferences for meat attributes such as tenderness, flavor, and appearance drive market demand. - Understanding consumer preferences guides breeding and production practices.

Meat quality is a complex interplay of various factors along the entire production and processing chain. By carefully managing these factors, producers can optimize meat quality and provide consumers with safe, flavorful, and nutritious meat products.

Meat Processing and Safety

Meat processing involves various techniques to transform raw meat into safe, palatable, and convenient products for consumption. Ensuring meat safety during processing is paramount to prevent foodborne illnesses and maintain product quality. Proper handling, sanitation, and adherence to regulatory standards are critical at every stage of meat processing. Here's an overview of meat processing and safety considerations:

1. Slaughtering and Dressing:

- Proper and humane slaughter methods minimize stress and ensure animal welfare.
- Immediate chilling prevents bacterial growth and reduces the risk of contamination.

2. Inspection and Grading:

- Government agencies perform inspection and grading to ensure meat safety and quality.
- Grading determines meat quality attributes like marbling and tenderness.

3. Meat Cutting and Processing:

- Butchers and processors cut meat into various cuts and products based on consumer demand.
- Hygienic practices and equipment sanitation prevent cross-contamination.

4. Marination and Flavoring:

- Marination and seasoning enhance flavor and juiciness.
- Marinades should be prepared and stored under safe conditions to prevent bacterial growth.

5. Cooking and Heating:

- Proper cooking temperatures ensure meat is safely cooked to kill harmful bacteria.
- Use a food thermometer to verify internal temperatures.

6. Packaging and Preservation:

- Packaging prevents contamination and maintains freshness.
- Vacuum sealing and modified atmosphere packaging extend shelf life.

7. Food Safety Measures:

- HACCP (Hazard Analysis and Critical Control Points) plans identify and mitigate potential hazards.
- Implement sanitation procedures to prevent cross-contamination and bacterial growth.

8. Temperature Control:

- Maintain appropriate temperatures during processing, storage, and transportation to prevent bacterial growth.
- Temperature abuse can lead to rapid bacterial multiplication.

9. Hygiene Practices:

- Implement strict hygiene practices, including hand washing, wearing appropriate protective clothing, and maintaining clean work surfaces.

10. Equipment Maintenance: - Regularly clean and sanitize equipment to prevent contamination and maintain product quality.

11. Allergen Management: - Proper labeling and separation of allergenic ingredients prevent cross-contact.

12. Traceability and Recall Systems: - Implement traceability systems to track the origin of meat products and facilitate recalls if necessary.

13. Compliance with Regulations: - Adhere to food safety regulations and standards set by government agencies.

14. Training and Education: - Properly train staff on food safety practices and provide ongoing education.

Ensuring meat safety requires a comprehensive approach that involves the entire production and processing chain. By adhering to strict safety protocols, maintaining hygiene, and prioritizing consumer health, meat processors can produce safe and high-quality meat products that meet consumer expectations and regulatory standards.

Slaughter Techniques

Slaughter techniques are methods used to humanely and efficiently convert live animals into meat for consumption. Proper slaughter techniques are critical to ensure animal welfare, maintain meat quality, and minimize stress for both animals and handlers. Slaughter techniques can vary based on the species, cultural practices, and regulatory requirements. Here are some common slaughter techniques:

1. Stunning:

- Stunning is the initial step in the slaughter process and is aimed at rendering the animal unconscious before proceeding to slaughter.
- Stunning methods include electrical stunning, captive bolt stunning, gas stunning, and mechanical stunning.
- Effective stunning prevents pain and stress for the animal and improves the overall quality of the meat.

2. Bleeding:

- After stunning, the animal is bled to remove blood from the body, which improves meat quality and reduces bacterial contamination.
- Proper bleeding ensures that the animal is insensible to pain and prevents consciousness during further processing.

3. Dressing and Evisceration:

- Dressing involves removing the hide or feathers from the carcass.

- Evisceration includes removing the internal organs from the carcass.
- Proper evisceration prevents contamination of the meat with intestinal contents.

4. Splitting and Chilling:

- After evisceration, carcasses are often split into halves for further processing.
- Carcasses are then chilled to reduce temperature and inhibit bacterial growth.

5. Halal and Kosher Slaughter:

- Halal slaughter follows Islamic dietary laws and requires the animal to be slaughtered by a trained person using a sharp knife while reciting a prayer.
- Kosher slaughter follows Jewish dietary laws and involves a specific method of slaughter and inspection.

6. Ritual Slaughter:

- Certain cultural and religious practices may involve specific ritual slaughter methods, often performed by trained individuals.

7. Mobile Slaughter Units:

- Mobile slaughter units provide on-site slaughter services, reducing animal stress caused by transportation.

8. Compliance with Regulations:

- Slaughter practices must comply with government regulations and standards to ensure food safety and animal welfare.

9. Animal Welfare Considerations:

- Slaughter techniques should prioritize animal welfare

by minimizing stress, pain, and fear during the process.

10. Training and Certification: - Handlers and slaughter personnel should be properly trained and certified in humane handling and slaughter techniques.

Slaughter techniques play a significant role in ensuring the safety, quality, and ethical treatment of animals destined for meat production. Proper stunning, bleeding, and evisceration are essential to producing meat products that meet food safety standards and consumer expectations. Additionally, adherence to cultural and religious practices, as well as regulatory requirements, is crucial in providing diverse consumer options while maintaining high standards of animal welfare and meat quality.

Meat Inspection and Food Safety Regulations

Meat inspection and food safety regulations are critical components of ensuring the safety and quality of meat products for human consumption. These regulations are established by government agencies to prevent the spread of foodborne illnesses, protect public health, and maintain consumer confidence in the meat industry. Here's an overview of meat inspection and food safety regulations:

1. Regulatory Authorities:

- Regulatory agencies responsible for meat inspection and food safety vary by country. In the United States, the U.S. Department of Agriculture (USDA) oversees meat inspection through the Food Safety and Inspection Service (FSIS). Other countries have similar agencies.

2. Mandatory Inspection:

- Meat products intended for human consumption typically undergo mandatory inspection by government-authorized inspectors.
- Inspection ensures that meat products are safe, properly labeled, and comply with regulatory standards.

3. Pre-Slaughter Inspection:

- Inspectors examine animals before slaughter to

identify signs of disease, injury, or other health issues.
- Animals showing signs of illness or conditions that may affect meat quality are condemned and not allowed for human consumption.

4. Post-Slaughter Inspection:

- After slaughter, carcasses and organs are inspected to identify any abnormalities, contamination, or disease.
- Carcasses and meat products that pass inspection are stamped or labeled as approved for consumption.

5. HACCP and Food Safety Plans:

- HACCP (Hazard Analysis and Critical Control Points) plans are a systematic approach to food safety that identifies and mitigates potential hazards at various stages of meat production.
- Meat processing establishments are required to develop and implement HACCP plans.

6. Sanitary Standards:

- Regulations establish sanitary standards for slaughterhouses and processing facilities to prevent cross-contamination, maintain hygiene, and ensure safe working conditions.

7. Labeling Requirements:

- Labels on meat products must accurately represent the product's contents, ingredients, origin, and proper handling instructions.
- Allergen information, nutritional labeling, and proper use of terms like "organic" and "grass-fed" are regulated.

8. Traceability and Recall Systems:

- Regulations often require establishments to maintain

records that allow for traceability of meat products from farm to consumer.

. Effective recall systems enable the prompt removal of unsafe products from the market if needed.

9. Import and Export Regulations:

. Meat imports and exports are subject to regulatory requirements to ensure that products meet safety standards and prevent the spread of diseases.

10. Compliance and Enforcement: - Regulatory agencies conduct routine inspections to ensure compliance with food safety standards. - Non-compliant establishments may face penalties, suspension of operations, or recalls.

11. International Standards: - International organizations such as the World Health Organization (WHO) and Codex Alimentarius establish global food safety standards that influence national regulations.

12. Consumer Education: - Regulatory agencies and industry stakeholders often collaborate to educate consumers about safe handling, cooking, and storage of meat products.

Meat inspection and food safety regulations play a vital role in protecting public health, preventing foodborne illnesses, and maintaining consumer trust in the meat industry. Compliance with these regulations ensures that meat products are safe, wholesome, and accurately labeled, contributing to a safe and reliable food supply chain.

Animal Nutrition Fundamentals

Animal nutrition is the science of understanding the dietary needs of animals to ensure their growth, health, reproduction, and overall well-being. Proper nutrition is essential for optimizing animal production, whether it's for meat, milk, eggs, or other products. Animal nutrition encompasses a range of concepts, including nutrient requirements, feed sources, digestion, and dietary management. Here are some fundamental aspects of animal nutrition:

1. Nutrient Categories:

- Animals require various nutrients for growth, energy, maintenance, and specific physiological functions.
- Essential nutrients include carbohydrates, proteins, fats, vitamins, minerals, and water.

2. Nutrient Requirements:

- Animals have specific nutrient requirements based on their species, age, weight, activity level, and production stage.
- Nutrient requirements are quantified in terms of recommended daily allowances (RDAs) or other similar measures.

3. Energy Sources:

- Carbohydrates and fats are primary energy sources for animals.
- Energy is essential for metabolic processes, growth, and physical activities.

4. Protein Needs:

- Protein is crucial for building and repairing tissues, enzymes, hormones, and immune function.
- Protein quality and amino acid composition are important factors in meeting animal protein needs.

5. Vitamins and Minerals:

- Vitamins and minerals play vital roles in various physiological processes.
- Each species has specific vitamin and mineral requirements.

6. Digestion and Absorption:

- Digestive processes break down complex nutrients into simpler forms for absorption.
- Ruminants, non-ruminants, and monogastric animals have different digestive systems and nutrient utilization.

7. Feed Ingredients:

- Feed ingredients include grains, forages, protein sources, and supplements.
- Formulating balanced diets involves combining different feed ingredients to meet nutrient requirements.

8. Feed Additives:

- Feed additives may include probiotics, prebiotics, enzymes, and growth promotants.
- These additives enhance nutrient absorption, gut health, and overall performance.

9. Feed Formulation:

- Feed formulation is the process of creating diets that

meet animal nutrient requirements.
- Balanced diets ensure optimal growth, reproduction, and overall health.

10. Water: - Adequate water intake is essential for digestion, temperature regulation, and overall metabolic functions.

11. Feeding Management: - Proper feeding practices include portion control, frequency, and timing of feeding. - Overfeeding and underfeeding can have negative impacts on animal health and production.

12. Nutritional Disorders: - Nutritional deficiencies or imbalances can lead to health issues, reduced production, and poor performance.

13. Special Diets: - Animals with specific needs, such as pregnant and lactating females, growing animals, or animals with medical conditions, may require tailored diets.

14. Sustainability and Environmental Impact: - Sustainable animal nutrition considers efficient resource use, minimizing waste, and reducing environmental impact.

Understanding animal nutrition is essential for ensuring the health and productivity of animals in various agricultural and livestock systems. By providing animals with well-balanced diets that meet their nutrient requirements, producers can optimize growth, reproduction, and overall well-being while contributing to sustainable and responsible animal production practices.

Nutrient Requirements for Different Species

Nutrient requirements for different animal species vary based on their physiological characteristics, growth stage, production goals, and environmental conditions. Providing animals with diets that meet their specific nutrient needs is essential for optimal growth, reproduction, and overall health. Here's an overview of nutrient requirements for some common animal species:

1. Cattle:

- Beef Cattle: Nutrient requirements vary with age, weight, and production stage (maintenance, growth, finishing).
    - Energy: Carbohydrates from grains or forages provide energy.
    - Protein: Adequate protein for muscle development and maintenance.
    - Minerals: Calcium, phosphorus, and trace minerals are crucial.
    - Vitamins: Vitamin A, D, and E are important.
- Dairy Cattle: Higher energy and protein requirements due to milk production.
    - Protein: High-quality protein for milk production and maintenance.
    - Energy: High-energy diets for milk production and lactation.

2. Poultry:

- Broiler Chickens: High-energy and protein diets for rapid growth.
  - Protein: High-quality protein for muscle development.
  - Energy: Carbohydrates and fats for energy.
  - Vitamins and Minerals: Including vitamin D, calcium, and phosphorus.
- Layers (Egg-laying Hens): High calcium for eggshell formation.
  - Calcium: For eggshell strength.
  - Protein: Adequate protein for egg production.
  - Vitamins: Vitamin D and E for egg quality.

3. Swine (Pigs):

- Growing Pigs: Balanced diets for growth and muscle development.
  - Protein: Quality protein for muscle growth.
  - Energy: Carbohydrates and fats for energy.
  - Minerals: Including calcium, phosphorus, and trace minerals.
- Gestating and Lactating Sows: Nutrient-dense diets for reproductive success.
  - Energy: High-energy diets during lactation.
  - Protein: Adequate protein for milk production.

4. Sheep and Goats:

- Sheep: Balanced diets for growth and wool production.
  - Protein: Adequate protein for muscle and wool growth.
  - Energy: Carbohydrates from forages.
- Goats: Nutrient-dense diets for meat and milk production.
  - Energy: High-energy diets during lactation.
  - Protein: High-quality protein for milk production.

5. Horses:

- Maintenance: Balanced diet for maintenance and health.
    - Fiber: Adequate fiber from forages.
- Performance Horses: Nutrient-dense diets for physical activity.
    - Protein: Adequate protein for muscle function.
    - Energy: Carbohydrates and fats for energy.

6. Fish (Aquaculture):

- Different fish species have varying nutrient requirements.
- Protein: High-quality protein for growth.
- Lipids: Fats for energy and essential fatty acids.
- Minerals: Including phosphorus and trace minerals.

7. Rabbits:

- Balanced diets for growth, reproduction, and maintenance.
- Fiber: Adequate fiber for digestion.
- Protein: Quality protein for growth and reproduction.

8. Dogs and Cats:

- Dogs and cats have specific nutrient requirements based on age, size, and activity level.
- Protein: High-quality protein for muscle maintenance.
- Fat: Essential fatty acids and energy.
- Minerals and Vitamins: Including calcium, phosphorus, and taurine for cats.

Note: Nutrient requirements are subject to change based on ongoing research and advancements in animal nutrition science. Consulting with animal nutritionists, veterinarians, and using up-to-date feeding guidelines is recommended to ensure animals receive appropriate and balanced diets tailored to their individual needs and production goals.

Feed Composition and Digestion

Feed composition and digestion play a crucial role in animal nutrition. Understanding the composition of various feed sources and how animals digest and utilize nutrients from those feeds is essential for providing balanced diets that meet their nutritional requirements. Here's an overview of feed composition, digestion, and nutrient utilization:

1. Feed Composition:

- Feed ingredients vary in their nutrient content, including carbohydrates, proteins, fats, vitamins, minerals, and water.
- Common feed ingredients include grains, forages, protein sources (like soybean meal), and additives.

2. Carbohydrates:

- Carbohydrates provide energy for animals and include sugars, starches, and dietary fiber.
- Monogastric animals (pigs, poultry) can digest starches, while ruminants (cattle, sheep) can ferment fiber through microbial digestion.

3. Proteins:

- Proteins are composed of amino acids and are essential for growth, tissue repair, enzyme function, and hormone production.
- High-quality protein sources provide essential amino acids.

4. Fats and Lipids:

- Fats provide concentrated energy and essential fatty acids.
- Lipids play a role in absorption of fat-soluble vitamins.

5. Vitamins and Minerals:

- Vitamins and minerals are required for various physiological functions, including bone health, immunity, and enzyme activities.

6. Water:

- Water is essential for digestion, nutrient absorption, temperature regulation, and overall metabolic functions.

7. Digestive Systems:

- Monogastric Animals (Non-Ruminants): Have a simple stomach and single-chambered stomach.
- Ruminants (Cattle, Sheep, Goats): Have a complex stomach with four compartments (rumen, reticulum, omasum, and abomasum).

8. Digestion in Monogastric Animals:

- Carbohydrates: Starches and sugars are broken down into glucose.
- Proteins: Proteins are broken down into amino acids.
- Fats: Fats are broken down into fatty acids and glycerol.
- Vitamins and minerals are absorbed in the small intestine.

9. Digestion in Ruminants:

- Rumen Fermentation: Microbes in the rumen break down fiber and produce volatile fatty acids (VFAs) as energy sources.
- Microbial Protein Synthesis: Microbes in the rumen

synthesize protein from nitrogen sources and amino acids.

- Omasum and Abomasum: Further digestion of nutrients occurs in these compartments.

10. Nutrient Absorption: - Nutrient absorption occurs mainly in the small intestine. - Nutrients are transported through the bloodstream to various body tissues.

11. Microbial Fermentation: - Microbes in the digestive system of some animals (like ruminants and certain fish) help break down complex carbohydrates and synthesize nutrients.

12. Feed Efficiency: - Feed efficiency measures how efficiently animals convert feed into products (e.g., weight gain, milk production).

13. Balancing Diets: - Formulating balanced diets involves combining different feed ingredients to meet nutrient requirements.

Understanding feed composition, digestion, and nutrient utilization is essential for designing diets that provide animals with the necessary nutrients for growth, reproduction, and overall health. By tailoring diets to the specific nutritional needs of each species, producers can optimize animal performance and contribute to sustainable and responsible animal production practices.

Feed Formulation and Ration Balancing

Feed formulation and ration balancing are critical processes in animal nutrition that involve creating balanced diets that meet the specific nutritional requirements of animals. Proper feed formulation ensures that animals receive the right amount of energy, protein, vitamins, minerals, and other nutrients necessary for growth, reproduction, and overall health. Here's an overview of feed formulation and ration balancing:

1. Nutrient Requirements:

- Determine the nutrient requirements of animals based on species, age, weight, production stage, and activity level.
- Nutrient requirements include energy, protein, amino acids, vitamins, minerals, and more.

2. Feed Ingredients:

- Identify available feed ingredients with known nutrient composition.
- Feed ingredients can include grains, forages, protein sources, vitamins, and minerals.

3. Nutrient Content Analysis:

- Analyze feed ingredient nutrient content to determine the amount of nutrients they provide per unit weight.

4. Energy and Protein Sources:

- Select energy-rich ingredients (grains, fats) and protein sources (soybean meal, fishmeal) to meet energy and protein requirements.

5. Amino Acid Balancing:

- Amino acids are building blocks of proteins.
- Formulate diets to provide the required amounts of essential amino acids.

6. Micronutrients:

- Include vitamins and minerals in appropriate amounts to ensure overall health and specific functions.

7. Cost Considerations:

- Balancing diets involves achieving nutrient requirements while minimizing costs.
- Some ingredients may be more cost-effective sources of nutrients.

8. Ration Balancing:

- Combine different feed ingredients to create a balanced ration that meets nutrient requirements.
- Rations may be formulated for specific production goals, such as growth, milk production, or egg laying.

9. Computerized Software:

- Computerized feed formulation software helps nutritionists calculate and optimize rations.
- Software considers nutrient requirements, ingredient availability, and cost factors.

10. Quality Control: - Regularly analyze feed samples to ensure that actual nutrient content matches formulated values. - Adjust formulations based on quality control results.

11. Monogastric vs. Ruminant Diets: - Ruminant diets consider microbial fermentation and utilization of fibrous materials. - Monogastric diets focus on efficient nutrient absorption in the small intestine.

12. Seasonal and Environmental Factors: - Environmental conditions, temperature, and animal stress may affect nutrient requirements.

13. Feeding Programs: - Develop feeding programs that transition animals from one diet to another as they progress through different growth stages.

14. Nutrition Consulting: - Nutritionists and animal scientists work together to formulate and adjust diets based on scientific knowledge and practical experience.

Effective feed formulation and ration balancing contribute to efficient animal production, optimal growth, reproductive success, and overall health. By carefully considering nutrient requirements, ingredient availability, and cost factors, producers can provide animals with diets that ensure their well-being while also enhancing production outcomes and sustainability.

Balancing Diets for Optimal Growth and Production

Balancing diets for optimal growth and production is a critical aspect of animal nutrition. Properly formulated diets ensure that animals receive the right nutrients in the right proportions to achieve their growth and production goals. Whether it's maximizing weight gain in livestock, enhancing milk production in dairy animals, or promoting egg production in poultry, balanced diets play a key role. Here's how to balance diets for optimal growth and production:

1. Define Production Goals:

- Clearly define the specific production goals, whether it's weight gain, milk production, egg production, or other outcomes.

2. Determine Nutrient Requirements:

- Consult with nutritionists and refer to scientific guidelines to determine the precise nutrient requirements for the target species, growth stage, and production goals.

3. Consider Energy Requirements:

- Choose energy-rich feed ingredients like grains, fats, and oils to meet the energy needs for growth and production.

4. Provide Adequate Protein:

- Incorporate high-quality protein sources to support muscle development, milk production, egg laying, and other physiological functions.

5. Amino Acid Balancing:

- Balance diets for essential amino acids to ensure proper protein synthesis and utilization.

6. Include Vitamins and Minerals:

- Include appropriate amounts of vitamins and minerals to support metabolic functions, bone health, and overall immunity.

7. Consider Fiber and Forages:

- For ruminant animals, provide adequate fiber and forages to support rumen health and microbial fermentation.

8. Account for Environmental Factors:

- Consider factors such as temperature, humidity, and stress that may affect nutrient requirements.

9. Optimize Feed Conversion Efficiency:

- Formulate diets to promote efficient feed conversion, ensuring that animals convert feed into desired products effectively.

10. Balance Nutrient Ratios: - Balance nutrient ratios to avoid over-supplementation or deficiencies, which can affect animal health and production.

11. Incorporate Growth Promotants (if applicable): - Growth promotants or additives can enhance growth and production efficiency, but their use should follow regulatory guidelines.

12. Monitor and Adjust: - Regularly monitor animal

performance, body condition, and production metrics. - Adjust diets based on actual outcomes and fine-tune formulations as needed.

13. Implement Feeding Programs: - Develop feeding programs that transition animals smoothly from one growth stage to another, adjusting nutrient levels accordingly.

14. Consider Sustainability: - Design diets that promote sustainable production practices, including efficient resource utilization and reduced environmental impact.

15. Seek Expert Guidance: - Collaborate with nutritionists, veterinarians, and animal scientists to ensure that diets are properly balanced and aligned with production goals.

Balancing diets for optimal growth and production requires a combination of scientific knowledge, practical experience, and attention to detail. By providing animals with well-formulated diets tailored to their specific needs, producers can achieve their production goals while ensuring animal health, well-being, and sustainability.

Incorporating Alternative Feeds and Byproducts

Incorporating alternative feeds and byproducts into animal diets is a cost-effective and sustainable approach to meeting the nutritional needs of animals while minimizing waste. Alternative feeds and byproducts refer to non-traditional feed sources, such as agricultural byproducts, food processing wastes, and unconventional ingredients, that can be used as feed ingredients. When properly evaluated and balanced, these alternative feeds can provide valuable nutrients to animals and contribute to efficient production. Here's how to incorporate alternative feeds and byproducts into animal diets:

1. Nutritional Analysis:

- Analyze the nutritional composition of alternative feeds and byproducts to determine their nutrient content and suitability for specific animal species.

2. Identify Nutrient Gaps:

- Identify the nutrient deficiencies or imbalances in the animal's current diet that can be supplemented by incorporating alternative feeds.

3. Feed Ingredient Characteristics:

- Consider factors like energy content, protein quality, fiber content, and digestibility of alternative feeds.

4. Compatibility and Palatability:

- Evaluate the palatability and acceptability of alternative feeds to ensure animals will consume them willingly.

5. Nutrient Balancing:

- Formulate diets that balance the nutrient profiles of conventional feeds and alternative feeds to meet animals' nutritional requirements.

6. Gradual Incorporation:

- Introduce alternative feeds gradually to allow animals to adapt to the new ingredients and prevent digestive disturbances.

7. Byproduct Utilization:

- Utilize byproducts from food processing industries, such as brewer's grains, distiller's grains, and fruit and vegetable pulps.

8. Local Availability:

- Consider locally available resources that may be underutilized or discarded, such as crop residues, agro-industrial byproducts, and surplus food.

9. Environmental Sustainability:

- Utilizing byproducts and alternative feeds can reduce waste and environmental impact by diverting materials from landfills.

10. Regulatory Considerations: - Ensure that the use of alternative feeds and byproducts complies with regulatory guidelines and safety standards.

11. Record Keeping: - Maintain records of the types and quantities of alternative feeds used in diets to track their impact on animal performance.

12. Monitor Animal Performance: - Regularly monitor animal growth, health, and production metrics to assess the effectiveness of alternative feeds in meeting nutritional needs.

13. Consult with Experts: - Seek advice from animal nutritionists, veterinarians, and extension specialists to ensure proper utilization of alternative feeds.

14. Adjust Formulations: - Be prepared to adjust feed formulations based on changes in feed availability, animal performance, and nutritional requirements.

15. Cost-Benefit Analysis: - Evaluate the cost-effectiveness of incorporating alternative feeds compared to traditional feeds.

Incorporating alternative feeds and byproducts into animal diets can help reduce feed costs, improve sustainability, and efficiently utilize available resources. However, careful evaluation, formulation, and monitoring are essential to ensure that the resulting diets meet animals' nutritional needs and support their growth, production, and overall well-being.

Apiculture: Beekeeping and Honey Production

Apiculture, commonly known as beekeeping, is the practice of raising and managing honeybees for the purpose of harvesting honey, beeswax, and other valuable bee products. Beekeeping has a long history and plays a crucial role in pollination, biodiversity, and the production of honey and other bee-related products. Here's an overview of apiculture, including beekeeping practices and honey production:

1. Hive Setup and Management:

- Beekeepers maintain hives, which are structured homes for honeybees. Common hive types include Langstroth, top-bar, and Warre hives.
- Hive components include frames, supers, brood chambers, and covers.

2. Bee Colony Structure:

- A colony consists of a queen bee, worker bees, and drones.
- The queen is responsible for laying eggs, while worker bees perform various tasks, including foraging, nursing, and hive maintenance.
- Drones are male bees whose primary role is to mate with virgin queens.

3. Beekeeping Equipment:

- Beekeepers use equipment such as bee suits, gloves,

smokers, and hive tools to manage hives safely and efficiently.

4. Hive Inspection:

- Regular hive inspections help monitor bee health, hive condition, and the presence of pests or diseases.
- Inspections involve examining frames for brood (eggs, larvae, pupae), honey storage, and overall colony strength.

5. Bee Foraging and Pollination:

- Bees forage for nectar, pollen, and water.
- Pollination services provided by bees are essential for many crop plants.

6. Honey Production:

- Bees collect nectar from flowers and transform it into honey through enzymatic processes and water evaporation.
- Beekeepers harvest honey by removing frames with capped honeycomb and extracting the honey using centrifugal extractors.

7. Beeswax and Propolis:

- Beeswax is produced by bees to construct honeycomb cells and store honey and pollen.
- Propolis, a resinous substance, is used by bees for hive sealing and protection.

8. Hive Health and Disease Management:

- Beekeepers monitor and manage hive health to prevent and address issues such as pests, parasites (like Varroa mites), and diseases (like American Foulbrood).

9. Swarm Management:

- Swarming is a natural behavior where a colony divides to form a new hive. Beekeepers may manage swarming to prevent hive loss.

10. Honey Extraction and Processing: - After harvesting honey, beekeepers extract it from honeycomb frames, filter it, and store it in clean containers. - Minimal processing preserves the natural qualities of honey.

11. Hive Products: - In addition to honey, beekeepers can harvest beeswax, pollen, royal jelly, and even venom.

12. Environmental Considerations: - Beekeeping practices should consider environmental sustainability, including providing forage and habitats for bees.

13. Education and Outreach: - Beekeepers often engage in education and outreach activities to promote beekeeping and raise awareness about the importance of bees.

Beekeeping is a rewarding endeavor that requires knowledge, patience, and commitment. Beyond honey production, beekeeping contributes to ecological balance through pollination and supports biodiversity. Beekeepers play a vital role in safeguarding the health of bee populations and ensuring the availability of bee products that are enjoyed by consumers worldwide.

Bee Anatomy and Behavior

Bee anatomy and behavior are fascinating aspects of the complex lives of honeybees and other bee species. Understanding the anatomy and behavior of bees is essential for beekeepers and researchers alike, as it provides insights into their roles within the hive, their interactions with the environment, and their contributions to pollination and honey production. Here's an overview of bee anatomy and behavior:

Bee Anatomy:

1. Exoskeleton:

- Bees have a hard exoskeleton made of chitin that protects their bodies.

2. Segmentation:

- The body is divided into three sections: head, thorax, and abdomen.

3. Head:

- The head contains compound eyes for detecting light and movement.
- Bees also have three simple eyes (ocelli) that detect light intensity.

4. Mouthparts:

- Bees have chewing mouthparts (mandibles) for manipulating wax and collecting pollen, and a proboscis for feeding on nectar.

5. Antennae:

- Antennae are sensory organs used for detecting odors and chemicals, communicating with other bees, and navigating.

6. Thorax:

- The thorax holds the wings and legs.
- Bees have four wings, two forewings and two hindwings, which they use for flight and temperature regulation.

7. Legs:

- Bees have six legs equipped with specialized structures for grooming, carrying pollen, and collecting nectar.

8. Abdomen:

- The abdomen contains vital organs including the digestive, reproductive, and respiratory systems.

Bee Behavior:

1. Foraging:

- Bees forage for nectar and pollen from flowers.
- Nectar is stored in their honey stomach and later converted into honey in the hive.
- Pollen is collected on their body hairs and used as protein-rich food for developing larvae.

2. Dance Language:

- Honeybees communicate the location of food sources using a dance language.
- The round dance indicates food sources nearby, while the waggle dance indicates distant food sources.

3. Hive Tasks:

- Worker bees perform various tasks such as nursing larvae, building honeycomb, foraging, guarding the hive, and processing nectar into honey.

4. Queen Bee:

- The queen bee is larger than workers and is responsible for laying eggs.
- She releases pheromones that regulate the behavior and development of other bees in the colony.

5. Worker Bees:

- Female worker bees are smaller and perform diverse roles within the hive.
- They transition through tasks such as cleaning cells, feeding larvae, building comb, guarding, and foraging.

6. Drone Bees:

- Male drone bees are larger and are raised for mating with virgin queens.
- Their primary function is reproductive, and they are expelled from the hive after mating.

7. Swarming:

- Swarming is a natural process where a colony divides to create a new hive.
- A swarm includes the old queen and a portion of the worker bees.

8. Hive Maintenance:

- Bees maintain hive temperature and humidity through fanning, evaporating water, and clustering.

Understanding bee anatomy and behavior is crucial for successful beekeeping, conservation efforts, and scientific research. It allows beekeepers to manage hives effectively, researchers to study bee populations, and everyone to appreciate the intricate and vital roles that bees play in pollination, ecosystems, and honey production.

Hive Management and Pollination

Hive management and pollination are two critical aspects of beekeeping that play a significant role in honey production, agricultural pollination, and the overall health of bee colonies. Effective hive management practices ensure the well-being of bee colonies, while pollination services provided by bees contribute to crop yield and ecosystem health. Here's an overview of hive management and pollination:

Hive Management:

1. Colony Inspection:

- Regular inspections of hives are essential to monitor colony health, check for diseases and pests, and assess the overall condition of the colony.

2. Hive Components:

- Beekeepers manage various hive components, including brood chambers, supers (for honey storage), frames, and covers.

3. Feeding and Supplementation:

- Beekeepers may provide supplemental feeding, especially during periods of nectar scarcity, to ensure colonies have enough food.

4. Disease and Pest Control:

- Implement integrated pest management (IPM) strategies to control Varroa mites, hive beetles, and other pests.

- Address disease outbreaks promptly and follow recommended treatment protocols.

5. Swarm Prevention and Management:

- Manage hive congestion and queen production to prevent swarming.
- If a swarm occurs, capture and rehive it to prevent colony loss.

6. Hive Hygiene:

- Keep hives clean and well-ventilated to prevent diseases and maintain a healthy environment.

7. Winterization:

- Prepare colonies for winter by ensuring they have sufficient food stores and protection from cold temperatures.

8. Queen Management:

- Monitor the performance of queen bees, replacing them if necessary to maintain colony productivity.

9. Splitting and Nucleus Colonies:

- Divide strong colonies to create nucleus colonies, which can be used for queen rearing or building new colonies.

10. Record Keeping: - Maintain accurate records of hive inspections, treatments, and hive health status.

Pollination:

1. Agricultural Pollination:

- Bees are essential pollinators for many agricultural crops, including fruits, vegetables, nuts, and oilseeds.
- Beekeepers offer hive rentals to farmers for pollination

services, enhancing crop yield and quality.

2. Biodiversity and Ecosystem Health:

- Bees contribute to the pollination of wild plants, supporting biodiversity and ecosystem health.

3. Pollinator-Friendly Practices:

- Encourage pollinator-friendly practices by avoiding pesticide use during bloom periods and providing forage resources.

4. Crop Synchronization:

- Timely placement of hives in fields with blooming crops ensures effective pollination.

5. Monitoring and Evaluation:

- Monitor hive strength, pollination efficacy, and crop yield to assess the success of pollination services.

6. Collaboration with Farmers:

- Collaborate with farmers to understand their crop requirements and customize pollination services accordingly.

7. Native Pollinators:

- Recognize the importance of native pollinators in addition to honeybees and support their conservation efforts.

Hive management and pollination are interconnected practices that contribute to successful beekeeping, sustainable agriculture, and ecosystem balance. By adopting effective hive management techniques and providing pollination services to agricultural partners, beekeepers play a vital role in supporting both the beekeeping industry and global food production.

Poultry Farming

Poultry farming, also known as aviculture, involves raising domesticated birds for their meat, eggs, and sometimes feathers. Poultry farming is a major agricultural industry that provides a significant source of animal protein for human consumption. It includes the raising of various bird species, including chickens, ducks, turkeys, and geese. Here's an overview of poultry farming:

1. Types of Poultry:

- Chickens: Raised for meat (broilers) and eggs (layers).
- Ducks: Raised for meat, eggs, and feathers.
- Turkeys: Primarily raised for meat.
- Geese: Raised for meat, feathers, and foie gras production.

2. Poultry Housing:

- Poultry housing varies based on the bird species and production purpose.
- Housing should provide shelter, protection from predators, proper ventilation, and adequate space.

3. Broilers (Meat Production):

- Broilers are chickens raised for meat production.
- Raised in controlled environments to optimize growth and feed efficiency.
- Harvested for meat at around 6-7 weeks of age.

4. Layers (Egg Production):

- Layers are chickens raised for egg production.
- Kept in layer houses with nesting boxes for egg-laying.
- Eggs are collected daily and processed for consumption or sale.

5. Hatcheries:

- Hatcheries produce day-old chicks for both broiler and layer production.
- Chicks are hatched from fertilized eggs in incubators.

6. Feeding and Nutrition:

- Proper nutrition is essential for growth, egg production, and overall health.
- Commercial feed is formulated to meet specific nutritional needs.

7. Disease Prevention and Biosecurity:

- Poultry farms implement biosecurity measures to prevent the spread of diseases.
- Vaccination and hygiene practices help maintain flock health.

8. Egg Collection and Grading:

- Eggs are collected, cleaned, and graded based on quality.
- Graded eggs are packaged for sale.

9. Meat Processing:

- Processing plants slaughter and process poultry for meat consumption.
- The process involves cleaning, evisceration, cutting, packaging, and distribution.

10. Alternative Poultry Production: - Free-range and pasture-raised systems allow birds outdoor access. - Organic poultry farming follows strict organic standards.

11. Environmental Considerations: - Poultry farming generates waste that must be managed properly to minimize environmental impact.

12. Market Demand: - Poultry products are in high demand due to their affordability and versatility.

13. Animal Welfare: - Ethical considerations include providing birds with proper care, space, and minimizing stress.

14. Research and Innovation: - Advances in genetics, nutrition, and management techniques improve poultry production efficiency.

15. Consumer Health and Safety: - Poultry farming practices impact the quality and safety of meat and eggs consumed by humans.

Poultry farming is a dynamic industry that contributes to global food security and provides livelihoods for many people. Proper management practices, biosecurity, and attention to animal welfare are crucial for sustainable and ethical poultry production.

Chicken and Turkey Production

Chicken and turkey production are two of the most significant segments within the poultry industry, providing meat for human consumption. Both chicken and turkey are popular sources of protein due to their versatility, nutritional value, and relatively lower cost compared to other meat sources. Here's an overview of chicken and turkey production:

Chicken Production:

1. Broilers:

- Broilers are chickens raised for meat production.
- They are raised in controlled environments to achieve rapid growth and efficient feed conversion.
- Marketed for meat consumption at around 6-7 weeks of age.

2. Housing and Management:

- Broiler houses are designed for optimal temperature, ventilation, and lighting conditions.
- Birds are provided with proper nutrition, clean water, and disease prevention measures.

3. Breeds and Genetics:

- Different broiler breeds and genetic lines are selected for traits such as growth rate and meat quality.

4. Feeding and Nutrition:

- Broilers are fed specially formulated diets to ensure rapid growth and meat quality.

5. Meat Processing:

- Processing plants slaughter, clean, and process broilers into various cuts and products for consumption.

6. Value-Added Products:

- Chicken meat is processed into a variety of value-added products, including nuggets, sausages, and ready-to-eat meals.

7. Environmental Considerations:

- Sustainable broiler production involves managing waste and minimizing environmental impact.

Turkey Production:

1. Turkey Farming:

- Turkeys are raised primarily for meat production.
- Turkey production involves similar practices as broiler production but with longer growth periods.

2. Housing and Management:

- Turkey houses are designed to accommodate the larger size of turkeys.
- Proper nutrition, water supply, and disease prevention are essential for turkey health.

3. Breeds and Genetics:

- Different turkey breeds are selected for meat production and traits such as meat yield and feed efficiency.

4. Feeding and Nutrition:

- Turkeys are fed specialized diets to support their growth and meat quality.

5. Meat Processing:

- Turkey processing involves slaughtering, cleaning, and processing turkeys into various cuts and products.

6. Holiday Demand:

- Turkeys are popular during holidays like Thanksgiving and Christmas, leading to seasonal demand.

7. Value-Added Products:

- Turkey meat is used in various products, including deli slices, ground turkey, and sausages.

8. Environmental Considerations:

- Sustainable turkey production focuses on waste management and environmental stewardship.

Chicken and turkey production are essential components of global food systems, providing affordable and accessible sources of protein. Sustainable production practices, animal welfare considerations, and adherence to safety and quality standards are crucial to ensuring the availability of safe and nutritious chicken and turkey products for consumers.

Egg Production and Hatcheries

Egg production and hatcheries are integral components of the poultry industry, specifically focused on the production of eggs for human consumption and the hatching of eggs to produce chicks for meat or egg production. Egg production involves raising hens specifically for egg laying, while hatcheries play a crucial role in producing day-old chicks that will later be raised for various purposes. Here's an overview of egg production and hatcheries:

Egg Production:

1. Layer Hens:

- Layer hens are chickens bred and raised for the primary purpose of egg production.
- Different breeds are selected for their ability to lay a large number of eggs.

2. Housing and Management:

- Layer houses provide comfortable conditions for hens to lay eggs in nesting boxes.
- Proper ventilation, lighting, and feed/water systems are essential for optimal egg production.

3. Egg Collection:

- Eggs are collected daily from nesting boxes.
- Workers ensure the eggs are clean, undamaged, and suitable for consumption.

4. Egg Quality and Grading:

- Eggs are graded based on size, shell quality, and cleanliness.
- Grading helps determine which eggs are sold as table eggs and which are used for processing.

5. Processing:

- Eggs may undergo washing, sanitizing, and packaging before distribution to consumers.

6. Value-Added Products:

- Some eggs are processed into value-added products such as liquid eggs, egg whites, and pre-cooked products.

7. Animal Welfare Considerations:

- Ethical egg production involves providing hens with adequate space, nesting areas, and access to the outdoors in some systems.

Hatcheries:

1. Incubation:

- Hatcheries use incubators to provide controlled temperature and humidity conditions for hatching eggs.

2. Egg Selection:

- Only fertilized eggs are selected for incubation, which are obtained from breeding flocks.

3. Hatching Process:

- During incubation, embryos develop inside the eggs.
- The hatching process typically takes around 21 days for chicken eggs.

4. Chick Processing:

- Once hatched, day-old chicks are sorted, vaccinated, and packaged for distribution.

5. Broilers vs. Layers:

- Some chicks are raised for meat (broilers), while others are raised as future egg-laying hens (layers).

6. Quality Control:

- Hatcheries maintain strict quality control measures to ensure healthy and viable chicks.

7. Genetic Improvement:

- Hatcheries often work in collaboration with breeding companies to produce genetically superior chicks for optimal production.

8. Biosecurity:

- Hatcheries implement biosecurity measures to prevent disease spread among chicks.

Egg production and hatcheries are essential components of the poultry industry, supporting the availability of eggs and meat to meet consumer demand. Ethical considerations, quality control, biosecurity, and sustainability play critical roles in ensuring the health and welfare of both hens and chicks throughout the production process.

The Future of Animal Husbandry: Sustainable Practices and Technological Innovations

The future of animal husbandry holds promise for sustainable practices and technological innovations that will enhance efficiency, animal welfare, and environmental stewardship. As the global population continues to grow and the demand for animal products increases, the need for responsible and efficient animal farming practices becomes paramount. Here's a glimpse into the future of animal husbandry:

1. Sustainable Farming Practices:

- Regenerative agriculture: Implementing practices that restore soil health, enhance biodiversity, and sequester carbon.
- Precision agriculture: Using data-driven technologies to optimize resource use, reduce waste, and minimize environmental impact.

2. Alternative Feeds:

- Utilizing alternative protein sources, such as insect-based feeds and algae, to reduce dependence on traditional feed ingredients.

3. Improved Animal Welfare:

- Enhanced housing and management systems that prioritize the well-being of animals and provide them

with more natural environments.
- Access to outdoor spaces for grazing and foraging in livestock systems.

4. Genetic Selection and Biotechnology:

- Further development of genetic selection methods to improve desired traits in animals, such as disease resistance and feed efficiency.
- Application of advanced biotechnology techniques, such as gene editing, to improve animal health and productivity.

5. Digital Technologies:

- Internet of Things (IoT) devices and sensors for real-time monitoring of animal health, behavior, and environmental conditions.
- Data analytics and predictive modeling for informed decision-making in animal management.

6. Vertical Farming and Indoor Agriculture:

- Advancements in vertical farming and controlled environment agriculture for efficient livestock and poultry production in urban areas.

7. Aquaculture Innovations:

- Sustainable aquaculture practices that reduce the environmental impact of seafood production.
- Incorporation of recirculating aquaculture systems and responsible fish farming techniques.

8. Reduction of Antibiotic Use:

- Continued efforts to reduce the use of antibiotics in animal production through better management practices, disease prevention, and alternative treatments.

9. Blockchain and Traceability:

- Implementing blockchain technology to enhance transparency and traceability in the food supply chain, ensuring product authenticity and safety.

10. Energy Efficiency and Renewable Resources: - Adoption of renewable energy sources, such as solar and wind, to power farms and reduce carbon emissions.

11. Integrated Farming Systems: - Implementation of integrated farming systems that combine different agricultural activities to optimize resource utilization and waste management.

12. Education and Collaboration: - Greater emphasis on educating farmers, consumers, and policymakers about sustainable and ethical animal husbandry practices.

The future of animal husbandry is shaped by a commitment to sustainability, innovation, and ethical considerations. By embracing technological advancements and adopting practices that prioritize animal welfare, environmental conservation, and human health, the agriculture industry can navigate the challenges of feeding a growing global population while preserving the planet for future generations.

Importance of Ethical and Environmental Considerations in Animal Husbandry

Ethical and environmental considerations are of paramount importance in animal husbandry, as they have far-reaching impacts on animal welfare, human health, biodiversity, and the sustainability of our planet. Integrating ethical practices and environmental stewardship into animal farming not only ensures responsible and compassionate treatment of animals but also contributes to a healthier ecosystem and a more secure food supply. Here's why ethical and environmental considerations are crucial in animal husbandry:

1. Animal Welfare:

- Ethical considerations prioritize the well-being of animals, ensuring that they are treated with respect, provided appropriate care, and live in conditions that allow them to express natural behaviors.

2. Moral Responsibility:

- As stewards of the environment and its inhabitants, humans have a moral obligation to treat animals with compassion and minimize their suffering.

3. Sustainable Agriculture:

- Ethical and sustainable animal husbandry practices contribute to long-term agricultural viability by preventing overexploitation of resources and maintaining ecosystem balance.

4. Human Health:

- Ethical and environmentally conscious practices can minimize the use of antibiotics and pesticides, reducing the risk of antibiotic resistance and pesticide residues in food products.

5. Biodiversity Conservation:

- Protecting ecosystems from the negative impacts of intensive farming helps preserve biodiversity and the habitats of other species.

6. Climate Change Mitigation:

- Sustainable animal farming practices help reduce greenhouse gas emissions, deforestation, and other activities that contribute to climate change.

7. Responsible Resource Use:

- Ethical and environmentally conscious animal farming minimizes the consumption of water, land, and energy resources, thus promoting efficient use and reducing waste.

8. Consumer Demand:

- Ethical and sustainable practices align with the growing consumer demand for ethically produced and environmentally friendly food products.

9. Reputation and Branding:

- Farms and companies that prioritize ethics and the environment often gain positive reputation and branding, attracting conscious consumers.

10. Legislation and Regulation: - Many countries have regulations and standards in place to ensure animal welfare and environmental protection in animal husbandry.

11. Long-Term Viability: - Ethical and environmentally responsible practices ensure that agricultural systems remain viable and productive for future generations.

12. Public Awareness: - Emphasizing ethical and environmental considerations helps raise public awareness about the interconnectedness of human, animal, and environmental well-being.

Balancing the needs of animals, people, and the environment is essential for the long-term sustainability of animal husbandry and the food supply chain. By embracing ethical and environmentally conscious practices, stakeholders in the agriculture industry can contribute to a more compassionate, resilient, and sustainable future for all.

Glossary of Key Terms

Here is a glossary of key terms related to animal husbandry:

1. **Animal Husbandry:** The science and practice of raising and caring for animals, including livestock, poultry, and other domesticated animals, for various purposes such as food, fiber, and labor.
2. **Livestock:** Domesticated animals raised for various purposes, including meat, milk, wool, and other products.
3. **Poultry:** Domesticated birds raised for meat and egg production, including chickens, turkeys, ducks, and geese.
4. **Breeding:** The controlled mating of animals to produce offspring with desired traits.
5. **Hatchery:** A facility where eggs are incubated and hatched to produce day-old chicks.
6. **Egg Production:** The process of raising hens for the purpose of collecting their eggs for consumption.
7. **Broilers:** Chickens raised for meat production, typically harvested at a young age.
8. **Layers:** Chickens raised for egg production, often kept in specialized housing with nesting boxes.
9. **Animal Welfare:** The ethical treatment and well-being of animals, ensuring they are free from unnecessary suffering and provided with appropriate care.
10. **Sustainable Agriculture:** Farming practices that prioritize long-term environmental, social, and economic sustainability.
11. **Biosecurity:** Measures taken to prevent the spread

of diseases among animals, often involving hygiene protocols and restricted access to farms.

12. **Genetic Selection:** The process of choosing animals with specific desirable traits to breed, aiming to improve the genetics of future generations.

13. **Precision Agriculture:** Using technology, data, and analytics to optimize agricultural practices, such as resource use, crop management, and animal husbandry.

14. **Regenerative Agriculture:** Farming practices that focus on improving soil health, biodiversity, and ecosystem function.

15. **Aquaculture:** The farming of aquatic organisms, such as fish, shellfish, and aquatic plants, in controlled environments.

16. **Biodiversity:** The variety of life forms in an ecosystem, including plants, animals, and microorganisms.

17. **Antibiotic Resistance:** The ability of bacteria to withstand the effects of antibiotics, often resulting from the overuse or misuse of antibiotics in animal husbandry.

18. **Traceability:** The ability to track the origin, journey, and processing of food products throughout the supply chain.

19. **Vertical Farming:** Growing crops or raising animals in vertically stacked layers, often in controlled environments like indoor farms.

20. **Integrated Farming:** Combining multiple agricultural activities, such as crop production, animal husbandry, and waste management, to optimize resource use.

21. **Carbon Sequestration:** The process of capturing and storing carbon dioxide from the atmosphere, often through activities like planting trees or improving soil health.

22. **Value-Added Products:** Processed food products that have undergone additional steps beyond basic processing, adding value to the product.

23. **Recirculating Aquaculture System (RAS):** A closed-loop system that recirculates water in aquaculture, allowing for efficient water use and waste management.

24. **Ethical Considerations:** Moral principles and values that guide responsible treatment and care of animals in farming practices.

25. **Environmental Stewardship:** Responsible management and conservation of natural resources to ensure the health and sustainability of ecosystems.

26. **Ecosystem Balance:** A state in which species and resources in an ecosystem are in harmony, contributing to the overall health and stability of the ecosystem.

27. **Internet of Things (IoT):** The network of interconnected devices and sensors that communicate and share data through the internet.

28. **Regulation:** Government-imposed rules and standards that govern various aspects of animal husbandry, including animal welfare and environmental protection.

29. **Responsible Resource Use:** Efficient utilization of natural resources, such as water, energy, and land, to minimize waste and environmental impact.

30. **Animal Welfare Certification:** Voluntary programs that certify farms and producers for adhering to specific animal welfare standards and practices.

31. **Sustainability Certification:** Programs that certify farms and producers for adopting sustainable agricultural practices and reducing environmental impact.

32. **Eco-friendly Practices:** Farming practices that prioritize environmental protection, conservation,

and minimal ecological impact.

33. **Carbon Footprint:** The amount of greenhouse gas emissions, particularly carbon dioxide, produced as a result of human activities.

34. **Animal Welfare Standards:** Guidelines and criteria established to ensure the well-being of animals in various farming practices.

35. **Food Security:** The availability and access to sufficient, safe, and nutritious food for all individuals.

36. **Beekeeping:** The practice of raising and managing bees for honey production and pollination services.

37. **Pasture-Raised:** Farming system where animals, such as poultry and livestock, have access to open pastures for grazing and foraging.

38. **Permaculture:** An approach to farming that emphasizes natural ecosystem design, sustainable practices, and self-sufficiency.

39. **Organic Farming:** Agricultural practices that avoid the use of synthetic chemicals, pesticides, and genetically modified organisms, focusing on natural and sustainable methods.

40. **Animal Welfare Regulations:** Laws and regulations that govern the treatment and care of animals in farming practices to ensure their well-being.

41. **Biotechnology:** The use of biological processes and living organisms to develop products and solutions, often used to improve animal health and productivity.

42. **Predator Control:** Measures taken to protect livestock from predators and minimize livestock losses due to predation.

43. **Food Safety:** Practices and regulations to ensure that food products are safe for consumption, free from contamination, and adhere to quality standards.

44. **Integrated Pest Management:** A sustainable

approach to managing pests by combining multiple strategies to minimize their impact on crops and animals.

45. **Community Supported Agriculture (CSA):** A model where consumers directly support farmers by purchasing shares of a farm's harvest in advance.

46. **Animal Welfare Advocacy:** Efforts to promote and raise awareness about responsible and compassionate treatment of animals in agricultural practices.

47. **Local Food Movement:** A movement encouraging consumers to buy food produced locally, supporting local farmers and reducing food transportation.

48. **Zoonotic Diseases:** Diseases that can be transmitted between animals and humans, highlighting the importance of animal health in preventing outbreaks.

49. **Ethical Farming Certification:** Certifications that verify that farms adhere to ethical and humane animal husbandry practices.

50. **Environmentally Friendly Labels:** Labels on food products indicating that the production methods used have minimal impact on the environment.

This glossary provides definitions for key terms related to animal husbandry, ethical considerations, and environmental practices in agriculture.

Recommended Resources and Further Reading

Here are some recommended resources and further reading materials for those interested in learning more about animal husbandry, ethical practices, and environmental considerations in agriculture:

Books:

1. "The Omnivore's Dilemma: A Natural History of Four Meals" by Michael Pollan
2. "Animal, Vegetable, Miracle: A Year of Food Life" by Barbara Kingsolver
3. "The Third Plate: Field Notes on the Future of Food" by Dan Barber
4. "The Meat Racket: The Secret Takeover of America's Food Business" by Christopher Leonard
5. "Defending Beef: The Case for Sustainable Meat Production" by Nicolette Hahn Niman
6. "Dirt to Soil: One Family's Journey into Regenerative Agriculture" by Gabe Brown
7. "The Soil Will Save Us: How Scientists, Farmers, and Foodies Are Healing the Soil to Save the Planet" by Kristin Ohlson
8. "The Humane Economy: How Innovators and Enlightened Consumers Are Transforming the Lives of Animals" by Wayne Pacelle

Websites and Organizations:

1. World Animal Protection (worldanimalprotection.org): An international organization focused on animal welfare and ethical treatment.
2. Sustainable Agriculture Research & Education (SARE) (sare.org): Offers resources on sustainable farming practices and research.
3. Humane Society of the United States (humanesociety.org): Advocates for animal welfare and responsible farming practices.
4. Regeneration International (regenerationinternational.org): Focuses on regenerative agriculture practices to improve soil health and ecosystems.
5. Animal Welfare Approved (animalwelfareapproved.org): Certifies farms that adhere to high animal welfare standards.
6. The Land Institute (landinstitute.org): Conducts research on perennial crops and sustainable agriculture.
7. Environmental Working Group (EWG) (ewg.org): Provides information on environmental impacts of food production.

Documentaries:

1. "Food, Inc."
2. "Cowspiracy: The Sustainability Secret"
3. "The True Cost"
4. "In Defense of Food"
5. "Rotten" (Netflix series exploring various aspects of the food industry)

Academic Journals:

1. Journal of Animal Science
2. Animal Welfare

3. Agriculture, Ecosystems & Environment
4. Frontiers in Sustainable Food Systems
5. Journal of Sustainable Agriculture

These resources offer a variety of perspectives and information on animal husbandry, ethical practices, and sustainable agriculture. Whether you're interested in learning about animal welfare, environmental impact, or responsible food production, these materials can provide valuable insights and knowledge.

www.ingramcontent.com/pod-product-compliance
Lightning Source LLC
Chambersburg PA
CBHW072206290526
45794CB00004B/1672